SPINY LOBSTERS

SPINY LOBSTERS
Through the eyes of the giant packhorse

John Booth

VICTORIA UNIVERSITY PRESS

TE WHARE WĀNANGA O TE ŪPOKO O TE IKA A MĀUI

VICTORIA UNIVERSITY PRESS
Victoria University of Wellington
PO Box 600 Wellington
victoria.ac.nz/vup/

Copyright © John Booth 2011

ISBN 978-0-86473-6550

This book is copyright. Apart from any fair
dealing for the purpose of private study, research, criticism
or review, as permitted under the Copyright Act, no part
may be reproduced by any process without the
permission of the publishers

National Library of New Zealand Cataloguing-in-Publication Data

Booth, John D. (John Duncan), 1948-
Spiny lobsters : through the eyes of the giant packhorse /
John Booth.
Includes bibliographical references and index.
ISBN 978-0-86473-655-0
1. Spiny lobsters. 2. Spiny lobster culture. 3. Jasus. I. Title.
595.384—dc22

Printed by 1010 Printing, China

Illustrations

Recruitment mechanism for packhorse spiny lobsters in New Zealand.	104
Recruitment mechanism for eastern rock lobsters in Australia.	105
Recruitment mechanisms among other spiny lobster species: red rock lobster *Jasus edwardsii*; ornate spiny lobster *Panulirus ornatus*; Japanese spiny lobster *Panulirus japonicus*; western rock lobster *Panulirus cygnus*; southern spiny lobster *Palinurus gilchristi*; and Natal spiny lobster *Palinurus delagoae*.	107
Places in northern New Zealand associated with the packhorse fishery.	152
World distribution of the main spiny lobster fisheries.	159
New Zealand red rock lobster CRA fishery areas.	167

TABLES

Earth's spiny lobster (family Palinuridae) genera: number of species in each genus, and depth and seafloor type of their typical habitat.	53
Geological eras, periods and epochs, from the emergence of the first lobsters in the Permian period.	56
Earth's spiny lobster species: scientific and common names	76–77
Estimated annual New Zealand commercial landings (whole weight) of packhorse, in tonnes, 1953–2008.	155

GRAPHS

Chart of evolutionary relatedness for the spiny lobster genera based on body characters. *After* Patek, S.N.; Feldmann, R.M.; Porter, M.; Tshudy, D. 2006: Phylogeny and evolution. *In: Lobsters: biology, management, aquaculture and fisheries.* Phillips, B.F. (ed) pp 113–145. Blackwell Publishing Ltd, Oxford *(with permission)*.	58
Size of female packhorse in typical spring-time catches from Spirits Bay and the Mercury Islands in the 1960s and the 1980s.	154
Estimated annual New Zealand commercial landings (whole weight) of packhorse, 1953–2008.	156
Jasus edwardsii puerulus settlement along the east coast of central New Zealand, 1978–2009.	168
Southern Oscillation Index, 1951–2011.	169
Trajectories of recruited biomass in 2001 for the CRA 3 (Gisborne) red rock lobster fishery based on model recruitments (left) and the Castlepoint settlement index (middle), and the resulting CPUE for the fishery (right).	184

CRA 7 (Otago) red rock lobster fishery CPUE plotted against the lagged local settlement index. — 184

DRAWINGS

Richard Pike's 1969 drawing of a large male *Sagmariasus verreauxi*. *After* Pike, R.B. 1969: A case study in research: crayfish. *In: Fisheries and New Zealand.* Slack, E.B. (ed) pp 95–108. Proceedings of a seminar on fisheries development in New Zealand. Department of University Extension, Victoria University of Wellington. — 43

The Foundation Seamounts rock lobster, the first new species of *Jasus* discovered in more than a century. *Rick Webber.* — 71

A generalised plan of the life cycle of shallow-water spiny lobsters. *Rick Webber.* — 86

Body parts of the final-instar packhorse phyllosoma and the puerulus into which it metamorphoses. *Rick Webber.* — 90

Identifying final-stage phyllosomas. *Rick Webber.* — 93

Distinguishing a spiny lobster puerulus from a shrimp, a crab megalopa, and a slipper lobster nisto. *Rick Webber.* — 95

Distinguishing the puerulus of the packhorse from the puerulus of the red rock lobster. *Rick Webber.* — 95

The external body parts of this male packhorse are typical of those of all spiny lobsters. *Rick Webber.* — 111

Mating spiny lobsters. *From:* Childress, M.J.; Jury, S.H. 2006: Behaviour. *In: Lobsters: biology, management, aquaculture and fisheries.* Phillips, B.F. (ed) pp 78–112. Blackwell Publishing Ltd, Oxford *(with permission).* — 131

Joseph Banks trading a piece of tapa cloth for a red rock lobster during James Cook's first voyage to New Zealand, on *Endeavour*, probably sketched by the Polynesian navigator Tupaia. *The British Library Board, Add. 15508, f.11.* — 148

Pesqueros used in the Cuban fishery for the Caribbean spiny lobster. *From:* Cruz, R.; Phillips, B.F. 2000: The artificial shelters (pesqueros) used for the spiny lobster (*Panulirus argus*) fisheries in Cuba. *In: Spiny lobsters: fisheries and culture.* Phillips, B.F.; Kittaka, J. (eds) pp 400–419. Fishing News Books, Oxford *(with permission).* — 162

Examples of collectors used to estimate levels of puerulus settlement: hogs-hair collector, the crevice collector, and three artificial seaweed collectors. *From* Phillips, B.F.; Booth, J.D. 1994: Design, use, and effectiveness of collectors for catching the puerulus stage of spiny lobsters. *Reviews in Fisheries Science 2*: 255–289 *(with permission).* — 179

Diagram of the set-up used by Jiro Kittaka to culture the phyllosomas of packhorse and four other spiny lobster species. *From* Kittaka, J. 2000: Culture of larval spiny lobsters. *In*: *Spiny Lobsters: fisheries and culture.* Phillips, B.F; Kittaka, J. (eds) pp 508–532. Fishing News Books, Oxford *(with permission).*	200
Artificial reefs used to enhance the abundance of Japanese spiny lobsters in Shizuoka on the east coast of Japan. *From: Man-made nursery for Japanese spiny lobsters in South Izu.* Shizuoka Prefecture, 1979. Brochure in Japanese.	209

PHOTOGRAPHS

Juvenile packhorse lobsters in a commercial pot. *John Booth.*	18
Red rock lobster phyllosomas on the author's palm. *Alan Blacklock, NIWA.*	20
The face of a spiny lobster. *Alan Blacklock, NIWA.*	22
Nat Davey correctly suspects this packhorse to be undersized. *John Booth.*	27
A catch taken over 2 days late in 1963 on the North Cape Shortie Patch by Cec Ruffell on FV *Florence. Cec Ruffell.*	28
Packhorse were tagged with the western rock lobster tag. *John Booth.*	30
T-bar tags are much more widely used these days. *Simon Anderson, Lat.37 (bottom left), Hallprint Fish Tags (bottom right).*	30
Distinguishing the sexes. *John Booth.*	32
Large trawl catches in south Westland in the early days. *Phil Prendergast.*	36
Bob Street discovered the migration of red rock lobsters in southern New Zealand, which in places takes the form of queues. *John Booth (left), Bob Street (right).*	37
The stranding of 1500 tonnes of Cape rock lobsters at Elands Bay, South Africa in 1997. Cover of *African Journal of Marine Science 28(2)* (2000). *Grant Pitcher, used with permission of the publishers NISC (Pty) Ltd.*	41
In juvenile packhorse the branchial regions on the sides of the carapace are far less developed than they are in the adult, and the lobsters are distinctly green. *Alan Blacklock, NIWA.*	44
Labels accompanying the packhorse of special interest at Te Papa. *John Booth.*	48, 49
Among Rick Webber's responsibilities at Te Papa is curating the holotype of *Sagmariasus verreauxi. John Booth.*	50
Renowned Dutch crustacean taxonomist Lipke Holthuis and his catalogue of the marine lobsters of the world. *Source unknown.*	51
The fossil *Sagmariasus flemingi* discovered in a marl pit near Nelson. *Marianna Terezow, GNS Science.*	54

Don Miller supposedly operating a radio from Maria Theresa Reef. *From Eade, J.V. 1976: Geological notes on the Southwest Pacific Basin in the area of Wachusett Reef and Maria Theresa Reef. NZOI Oceanographic Summary 9.*	64
The *Albatros'* engagement with the *Southern Raider* before she was shelled and sunk on 9 October 1986 in the south Indian Ocean. *John Chadderton.*	66
The St. Paul rock lobster was known only from around St. Paul Island and the nearby Amsterdam Island until John Chadderton potted them over a broad area of the central Indian Ocean. *John Chadderton.*	69
John Chadderton caught what appeared to be two distinct forms of *Jasus* in the Indian Ocean. *John Chadderton.*	70
The Cape jagged lobster is the deepest-living of all spiny lobsters and has most often been found associated with seamounts. *Ring of Fire Expedition.*	70
New Zealand's spiny lobsters. *All photos by Alan Blacklock, NIWA, except Panulirus femoristriga, by Tin Yam Chan.*	72
New Zealand's slipper lobsters. *Antipodarctus aoteanus (Darryl Torckler); Scyllarides haanii (John Booth, NIWA); Ibacus alticrenatus (Alan Blacklock, NIWA); Ibacus brucei (G. Millen, from Brown, D.E.; Holthuis, L.B. 1998: The Australian species of the genus Ibacus (Crustacea: Decapoda: Scyllaridae), with the description of a new species and addition of new records. Zoologische Mededelingen 72: 113–141, used with permission); Antarctus mawsoni and unidentified scyllarine (NORFANZ); Arctides antipodarum (Malcolm Francis).*	75
The plankton is sampled with a huge, fine-meshed net. *Steve Mercer, NIWA.*	85
Net sample containing mainly krill but also spiny lobsters. *Dick Singleton, NIWA.*	85
Richard Pike's drawings of the final-stage phyllosomas of the packhorse and of the red rock lobster. *Jean-Claude Stahl and Raymond Coory, Te Papa.*	86
Packhorse have the largest egg clutches of all spiny lobsters; recently extruded eggs; eggs close to hatching, showing the paired eyes and the red pigment spots on the legs; ruptured egg capsules in an egg mass at hatching. *Graeme Moss and Alan Blacklock, NIWA.*	88
A packhorse naupliosoma, first-instar phyllosoma, and late-instar phyllosoma. *Alison MacDiarmid and Graeme Moss, NIWA.*	89
Final-stage phyllosomas of the red rock lobster. *John Booth, NIWA.*	90
A red rock lobster puerulus immediately after metamorphosis. *Simon Anderson, Lat.37.*	90

Illustrations

A lone pohutukawa tree clinging to the eastern side of the extreme promontory of Cape Reinga marks the 'place of leaping'. *John Booth.*	96
Red rock lobster pueruli under a boulder on the intertidal shore of Castlepoint in July 1990. *John Booth, NIWA.*	108
The beach where this highly unusual intertidal settlement of pueruli takes place is just below the Castlepoint lighthouse on the open coast, directly above the gazer's eyes. *Lloyd Homer, GNS Science.*	108
Panulirus pascuensis, known from three remote spots in the South Pacific Ocean. *Alan Blacklock, NIWA.*	110
Underside of the anterior part of a female packhorse, showing the feeding appendages, mouth parts and excretory pore. *John Booth.*	113
Views from above of an immature male and a mature female packhorse with the carapace, shell of the top of the tail, and the upper musculature removed. *John Booth.*	114
The reddish-brown of the very young juvenile tends towards dark green in older juvenile packhorse. *John Booth (left); Institute for Marine and Antarctic Studies, Tasmania (right).*	122
Kina (sea urchin) barrens. *Steve Mercer, NIWA.*	125
A small red cod had consumed a puerulus and young juvenile red rock lobster. *John Booth, NIWA.*	126
Intact spermatophore, or tar spot, on a female western rock lobster. *John Booth, NIWA.*	132
Red rock lobster with hatching naupliosomas. *NIWA.*	133
Large male red rock lobster. *John McKoy, NIWA.*	134
Bacterial shell disease in packhorse and other spiny lobsters is often seen as 'burns' on the shell and erosion on the tail fan. Occasionally the tail erosion spreads so much that it becomes unsightly. *John Booth.*	136
Packhorse fishing vessel *Medea*'s home port is Russell in the Bay of Islands. *John Booth.*	138
The mesh-covered, steel-framed pots used for packhorse are larger than those used in most other spiny lobster fisheries. *John Booth*	141
Measuring the carapace length of a small packhorse before tagging it. *John Booth.*	142
Often on Bill Hopkins' fishing boats the packhorse were held in holding pots before being delivered to port. Bill Hopkins died at the age of 80. *Fraser MacLean (left) and John Booth (right).*	145

Red rock lobster pot or taruke koura. *From* Best, E. 1929: Fishing methods and devices of the Maori. *Dominion Museum Bulletin 12.*	149
Bag-net or pouraka, baited with a crab. *From* Hiroa, Te R. 1926: The Maori craft of netting. *Transactions and Proceedings of the New Zealand Institute 56:* 597–646.	149
Dover Samuels had a good-sized packhorse on display in his parliamentary office. *John Booth.*	150
Packhorse were briefly trawled in the Far North in the early 1960s. *Stuart McFarlane.*	161
Tangle nets are used to catch the Japanese spiny lobster and other closely related species in Asia. *Andrew Jeffs*	162
Arabian whip lobsters, a deep-water species that is both target and by-catch for trawlers in parts of the Arabian Sea and Gulf of Oman. *John McKoy, NIWA.*	162
John Coombes' 1975 dive catch near Whitianga, in Mercury Bay. *John Coombes.*	164
Brian Hoult caught this 6.3-kilogram packhorse on a hook while snapper fishing in Bream Bay in 2003. *The Northern Advocate.*	165
Minimum legal size for the red rock lobster in New Zealand is a tail-width measurement. *Alan Blacklock, NIWA.*	171
Enormous catches were made of the red rock lobster in the early days of the New Zealand fishery. *Source unknown.*	174
Hauling a packhorse pot on FV *Toiler. Bill Hopkins.*	176
The crevice collector used to derive indices of settlement of the red rock lobster in New Zealand and parts of Australia. *Jeff Forman, NIWA.*	179
Newly settled red rock lobster pueruli. *Alan Blacklock, NIWA.*	181
The red rock lobster became so abundant just a few years after Te Tapuwae o Rongokako Marine Reserve on the east coast of the North Island of New Zealand was established in 1999 that large individuals could be found foraging on exposed intertidal reef platforms during the day. *Kerry Fox (left) and Jamie Quirk, DOC (right).*	191
The mainly rocky shore of the Cape Rodney-Okakari Point Marine Reserve near Leigh in northeast New Zealand is visited by thousands upon thousands each year. *Glass Bottom Boat, Goat Island Marine Reserve.*	192
The surface of the water is turned white with the thrashing of kingfish in the pot. *John Booth.*	193
The leatherback turtle, finally freed of its ensnarement, is prepared for winching back over the gunwale. *John Booth, NIWA.*	194

Department of Conservation's Debbie Freeman measures a red rock lobster. *Jamie Quirk, DOC.*	195
Jiro Kittaka displays Rick Webber's line drawing of the final-stage phyllosoma of the red rock lobster just given to him by the author at the 7th International Conference and Workshop on Lobster Biology and Management. *Marney Penn.*	198
The set-up used by Jiro Kittaka to culture the phyllosomas of packhorse and four other spiny lobster species. *John Booth, NIWA.*	200
Packhorse puerulus. *Graeme Moss, NIWA.*	201
Late-stage (Instar 14) packhorse phyllosomas, pueruli and first-instar juveniles, and larger (300 grams) juveniles cultured in Tasmania. *Arthur Ritar, Tasmanian Aquaculture & Fisheries Institute.*	203
In Vietnam, the pueruli of tropical spiny lobsters are caught in fine-meshed nets and ongrown to market size in floating cages. *Andrew Jeffs.*	205
Red rock lobster pueruli taken on crevice collectors at Gisborne were ongrown in captivity for enhancement experiments. *John Booth, NIWA.*	207
Japanese spiny lobsters are highly revered for ceremonial occasions. *John Booth, NIWA.*	211
From earlier times: tailing rock lobsters. Those not tailed were boiled. *John Booth, NIWA.*	212
Man, two boys, and a pile of crayfish, at the back of Harry Daniel's fish shop, Manaia. *J.C. Hosie Collection, Alexander Turnbull Library, Wellington.*	214
Nat Davey keeps live his packhorse in fine condition in holding pots until he is ready to land them. *John Booth.*	215
At $90 per kilogram, not everyone will be able to afford these live packhorse at this Auckland fish market. *John Booth.*	215
Packing live red rock lobsters for export. *John Booth, NIWA.*	216
However you chose to enjoy your packhorse tail, cooked or raw, don't forget about all the goodies still within the body. *John Booth.*	219
A demoiselle examining a large packhorse spiny lobster. *Malcolm Francis.*	222

PREFACE

Well over a century ago a strange monster of a lobster was delivered late one afternoon to the doors of what was then the Colonial Museum in Wellington, New Zealand. The leviathan was examined soon after by the museum's Assistant, Thomas Kirk, who declared it to be a new species of spiny lobster, *Palinurus tumidus*. A male, its first legs were as thick as a child's wrist, and it was no less than 24 inches long from head to tail, with a girth of 21.5 inches. Thomas later pondered if he should have named it *giganteus*. He organised for the lobster to be preserved, and later someone varnished it in an attempt to prolong its shelf life. This enormous trussed carcass still resides—seldom visited—in Wellington, now in the collections of the Museum of New Zealand Te Papa Tongarewa. When I examined it in 2006 it was a little ragged around the edges but nevertheless still in reasonably good nick given how long it'd been there in its strong, custom-built cardboard box. And two things in particular caught my eye: both the location and the name labels had been altered. The originals said that this lobster had been collected at WHANGAROA, but an extra 'I' had been inserted by hand, so the label now read WHAINGAROA. And someone had changed its name to *Jasus verreauxi*.

In fact, this spiny lobster was not *Palinurus tumidus*, nor *Jasus verreauxi*, but the packhorse *Sagmariasus verreauxi*. It had first been described in the middle of the 19th century, *Palinurus tumidus* being just one of several strange scientific names it had picked up along the way. And its present name is an excellent title. *Sagmariasus* is a play on the Greek word for packhorse, *sagmarion*, for when large this lobster, with its bulging sides, bears a remarkable resemblance to a beast of burden.

Much of the field research into the packhorse spiny lobster, that archetypal monster so symbolic of the waters of New Zealand's 'winterless north', took place from the late 1960s to the early 1980s. Northland back then was characterised by small hamlets, a relatively undeveloped infrastructure with many of the roads unsealed, rather humble homes amid often weedy pastures, and pretty tough individuals, innovative and making do. There was not an awful lot of wealth to be seen, especially among the disproportionately high Māori population.

Packhorse constantly inveigled me back up to this northern part of New Zealand in those early decades. I had grown up in the region so I knew it well, and from time to time I found myself longing to be out of Wellington, fishing and fossicking and exploring those clear waters and beguiling shores of Northland. And learning more about packhorse. At the time I was working for what was then the Ministry of Agriculture and Fisheries, mainly on the much more commercially significant red spiny lobster (more commonly known as the red rock lobster), but I'd combine work on both species whenever I came north. I could never say that I particularly liked packhorse—they were hellish to handle—but, more, I was intrigued by them: prehistoric-looking, often huge and always powerful, hardy, and yet at times almost sinister.

In the chapters that follow I reflect on the many notable things that characterise the extraordinary lives of packhorse and their relatives. Heavily armoured, with few predators once they are large, spiny lobsters look like they should have died out with the dinosaurs. In fact they arose when *Tyrannosaurus* and its cousins reigned, which was about the same time that the first mammals appeared. From then on their home bases were relocated on the coattails of the juggernaut that is plate tectonics. And the packhorse, among the most primitive of all lobsters, shows no sign of being unable to cope with any of this. To the contrary, as the field results unfolded, packhorse revealed how exquisitely suited they are to this little corner of the world. By reaching a monstrous size, with cornucopian larval production, they have adapted well to living around this couple of remote islands. To maintain their foothold in New Zealand waters this remarkable lobster has had to become a great trekker. Each year many thousands of young packhorse trudge steadfastly and purposefully—in just one direction.

Packhorse can't claim much in the way of overseas experience: they are known from the shores of only New Zealand and Australia. The same cannot be said, though, of spiny lobsters in general: they live around most main land masses, as well as at some of the most remote places on Earth. In 1986, on the other side of the world from New Zealand, a fishing vessel was revealing remarkable new information on the distribution of a very close relative of packhorse, the St. Paul rock lobster—information that changed completely our understanding of the extent to which these spiny lobsters populate our planet. What happened next could have turned out to be one of the most shocking acts of piracy—and murder—at sea in peace-time 20th century. A superpower despatched to the seafloor a fishing vessel which had had aboard more than 20, including several New Zealanders.

In the late 1970s, I had found myself late one night taking a second and much closer look at an item that lay in the large black sorting tray before me. We were hove to 20 kilometres or so east of Gisborne, in the west part of the South Pacific Ocean off the east coast of New Zealand, over waters 200 metres deep. The fine-meshed midwater trawl had just been hauled in and its contents delivered to us on the deck below to sort. Amidst the dank odour of diesel and engine fumes rising up the stairwell from the deck below, and this deck's dull illumination, lay a host of shrimp-like animals and small fish. I had been sampling the oceanic distribution of the small postlarvae of spiny lobsters, lively transparent little replicates of the adult 25 millimetres or so in length. So far this sample had contained six. But what was particularly catching my eye now was a remarkably leggy animal, almost 50 millimetres long, leaf-like and crystal-clear. It was the first live phyllosoma I had ever seen. Bearing no resemblance at all to a spiny lobster, and yet the larva of one, this animal was clearly ideally suited to a planktonic existence. Not only was it next to invisible to its predators, its wide, flat shape was perfect for drifting on ocean currents. Placed in a bowl of fresh seawater it moved feebly, ineffectually flailing its legs in a hopeless—almost tired—bid to escape. This contrasted strongly with the much more vigorous postlarva it would have soon become had it not been ensnared in our net. Within minutes it sank to the bottom of the bowl and stopped moving altogether.

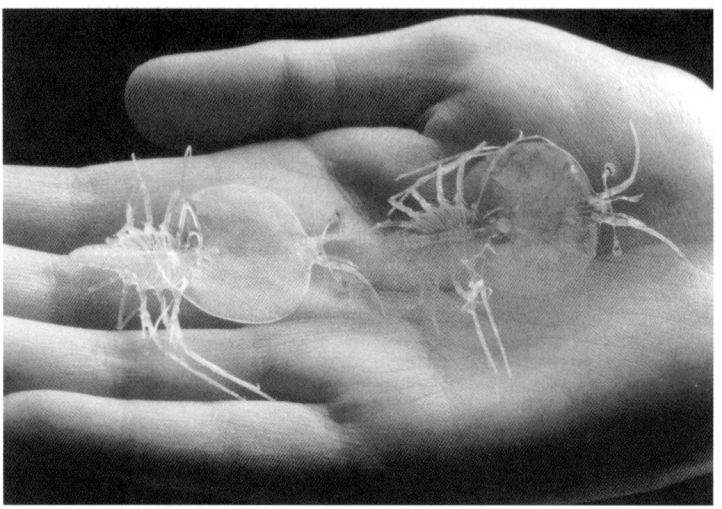

This phyllosoma larva belonged to another of the packhorse's cousins, the red rock lobster *Jasus edwardsii*—and it was already between one and two years old. Later in that voyage, as our sampling took us further and further from the coast and out into the vast expanses of the South Pacific Ocean, each of our plankton samples began to contain dozens, sometimes hundreds, of these remarkable larvae. For—almost unfathomably—it's out here in the open ocean, tens to hundreds of kilometres from land, that the larvae of all coastal spiny lobsters must spend many months if they are later to successfully take their place on the nearshore seabed. One of the most bewildering aspects of this oceanic phase is just how the larvae avoid being swept away altogether from their parent grounds, and how they eventually make it back to the coast from such distances.

Back on shore, packhorse take on more familiar appearance, going about their business of competing with their own and with others, growing, breeding.

Important questions concerning spiny lobster biology are now more tractable because the animals are close at hand, and because a lot of research has been directed at them due to their commercial value. One thing that had puzzled scientists was why spiny lobsters stored their urine in a bladder—they don't really need one as they can release waste directly into their environment as quickly as it is produced. It has been revealed that the accumulated contents of the bladder are critical to the manner in which lobster interact with each other, influencing all sorts of behaviour—combative, conciliatory, social, even amorous.

Most spiny lobsters that survive the perils of their time out in the open ocean, and the hazards of pre-adolescence, die at the hand of humankind. We've all seen them in fish markets and flinched at the cost—usually priced per 100 grams in an attempt to assuage the blow. Accordingly, the commercial exploitation of spiny lobsters is a very serious business. When I began to look into the catches of packhorse in New Zealand, stories abounded of catch rates so remarkable as to be almost unbelievable. And one name stood out above all others: Bill Hopkins. 'In the 1960s, I caught large quantities of packs in Spirits Bay in FV *Carol Ann* and later FV *Provider*. Lifts for one day peaked at 80 sacks or 4 tonnes,' Bill wrote of his early potting. It was not rare for him to land 7 tonnes in a two-day trip. 'It was a piss-poor day if there was only a tonne,' he told me. The large packhorse were so difficult to bag, invariably hooking the pointed tips of their legs into the weave of the sacks, that the crew got into the habit of leaving them for a while in the sun to quieten. In his record season he landed 112 tonnes.

Sustained heavy fishing has hit for six many spiny lobster stocks. Close to half of the world's most important fisheries display all the hallmarks of overfishing. More effective management intervention is desperately needed. But it is also now clear that spiny lobster stocks carry on lives of their own, reluctant to conform to the expectations of classical fisheries management theory. This is because natural events—including those associated with El Niños—can totally swamp the effects of human intervention by controlling the year-to-year levels of postlarval settlement and, in turn, fishery production. For reasons still to be disclosed, some New Zealand stock assessment experts have been unwilling to acknowledge this. Almost unfathomably, they chose to essentially ignore the widely available, systematically collected, fishery-independent data on settlement levels when assessing the status and prospects of our red rock lobster stocks.

One way of ameliorating the effects of overfishing on global spiny lobster production is through aquaculture. In 1987, Jiro Kittaka of Japan's Kitasato University became the first person ever to grow a spiny lobster species through all its larval stages: it was the Cape rock lobster *Jasus lalandii* and it took 10 months. He went on to grow four more species of spiny lobster to settlement, including both of New Zealand's commercial species, packhorse becoming his main focus. Larvae produced by his cultured packhorse hatched larvae of their own. And packhorse remains one of just a small handful of species of spiny lobster that are seen today as promising candidates for aquaculture.

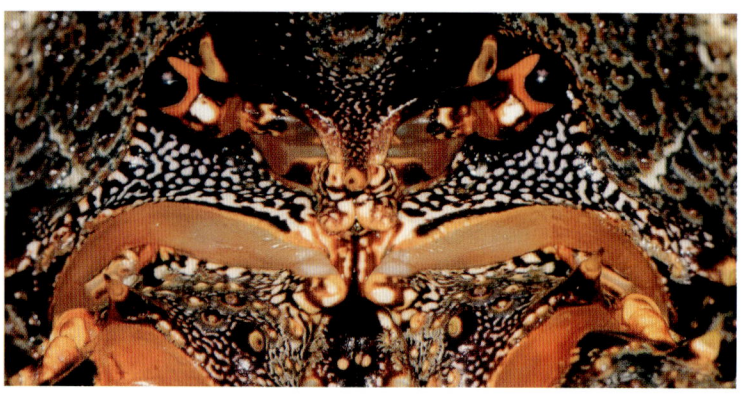

For the chapters that follow, the packhorse is my exemplar, from which I extend my enquiry to its *Jasus* relatives in particular and all other spiny lobsters in general. For the accounts of spiny lobster distribution, biology, ecology, aquaculture and fisheries, I have drawn heavily on the knowledge and skills of numerous colleagues. I have used results from their published research and reviews, and my writings have greatly benefited from both face-to-face and e-mail dialogue. And—perhaps most importantly—I have relied on their reviews and critiques of my writings.

To you all, for so generously giving of your time, expertise and assistance, my sincere thanks. I am especially grateful to the following people: Mike Beardsell, Mike Bradstock, Joe Cave, John Chadderton, Steve Chiswell, Sarah Creswell, Adam Davey, Nat Davey, Debbie Freeman, Ray George, Lester Goodfellow, Johan Groeneveld, Natalie Halley, Bill Hopkins, Bruce Hayward, Kristina

Hjelm, Andrew Jeffs, Chris Kennedy, Jiro Kittaka, Foss Leach, Fraser MacLean, Andy McKenzie, Reyn Naylor, Megan Oliver, Arthur Ritar, Don Robertson, Bob Street, Dover Samuels, Kevin Sullivan, Daryl Sykes, Rick Webber, Scott Westley and Serena Wilkens.

Special thanks to Kyleigh Hodgson of Victoria University Press for her excellent editor's eye; to Rick Webber for his top-notch drawings; and to Chris Kennedy for his keen sense of colour and proportion in map production.

Finally, I am grateful to Fergus Barrowman of Victoria University Press for taking on this book, and to the National Institute of Water and Atmospheric Research, the Department of Conservation, the Ministry of Fisheries, and the New Zealand Rock Lobster Industry Council for their generous contributions towards its publication.

Kia ora tatou.

CHAPTER 1

WIDE WANDERINGS

We're talking about a pretty remote place. Northland thrusts 100 kilometres up into the South Pacific Ocean beyond New Zealand's northernmost town of any size, Kaitaia. This narrow peninsula is heavily sanded on each side and, with much of it low-lying, is at imminent risk of deluge from a large tsunami. Gravel high on the dunes of Great Exhibition Bay, propelled 30 metres up there one particular day in the past several hundred years, attests to this menace.[1] The farthest part of this peninsula has remarkably varied topography. It was once a group of small offshore islands, which became linked to the mainland quarter of a million years ago by a sand tombolo that choked the strait.[2] Nowadays the 200-metre high points behind Cape Reinga and North Cape stand proud, overlooking the low-lying sweep south of Ninety Mile Beach in the west and Great Exhibition Bay in the east. And directly below North Cape, only a stone's throw from Surville Cliffs at the very tip of the country, is Parengarenga, New Zealand's northernmost harbour. On sunny days the pure-white silica sands of Kokota, the south head of Parengarenga Harbour, can be blinding. But the splendour of this inlet, its mangrove-fringed phalanges protruding west almost to the back dunes of Ninety Mile Beach, belies its danger. The entrance is narrow and shallow, a bar harbour that any boatie should take most seriously, and it has witnessed its share of tragedy (recently the *Busy Bee* and an accompanying vessel—16 thrown into the water, two perishing).[3]

Yet it was over this Parengarenga bar that fishermen in the 1960s would make their way early each morning to reach a small but very productive fishing ground nearby. Still others arrived from ports to the south. Waters about 100 metres deep a few kilometres southeast of North Cape were well known as the Shortie

Map of New Zealand showing places mentioned in the text. The inset is of the Far North.

Nat Davey correctly suspects this packhorse, large as it is, to be under-sized. For packhorse, the minimum legal size—the size below which the lobster cannot legally be retained—is 216 mm tail length. In most other spiny lobster fisheries, however, minimum legal size is based on carapace length, a much less flexible and more precisely measured body part.

Patch, because here large numbers of packhorse *Sagmariasus verreauxi*—small individuals, hence the name 'shortie'—were available, with possibly 100 tonnes being taken each season. So it continued until 1969, by which time it had become evident that the 6-inch tail-length size limit allowed few, if any, of the lobsters to breed before becoming available to the fishery. The minimum legal size was increased to 8.5 inches. Now that almost all lobsters on the Shortie Patch were undersized, most fishers went elsewhere, many closer to Cape Reinga where there were far greater proportions of large packhorse.[4]

It wasn't until the second packhorse tagged on the North Cape Shortie Patch had been captured far to the south that the first recapture became believable. After all, one tag recapture at a totally unexpected and distant place might simply be a prank, or a mistake. But now, in July 1979, a fisherman working out of the west coast port of New Plymouth reported to the local fisheries officer his catching of a tagged packhorse near Cape Egmont, close to where a tagged pack had been reported a few months previous.[5] A visit to this location, 500 kilometres south of Cape Reinga as the crow flies, reveals a shore not normally associated with packhorse: prevailing southwest swells and large andesite boulders, all overlooked by a conical peak that is snow-covered in winter. Not only, it seemed, was there westward movement of small packhorse from the Shortie Patch, into the main fishing grounds and breeding area near Cape Reinga, but also wanderings much, much further afield.

And greater surprises concerning the sheer extent of movements were to come.

φφφ

The Royal New Zealand Navy fisheries patrol vessel is already here, hauling the last of the pots. HMNZS *Pukaki*—swift, sinister, sleek, steel, just on 33 metres with a crew of 21—has aboard the Far North Fisheries Officer Fraser MacLean. This was March 1977 and he had learnt that there were dozens of illegal pots on the Shortie Patch which must now be confiscated. With no vessel capable of working this distant, open-water location at his disposal, the navy has taken on the job. We are on the much slower and more elderly 19-metre *Ikatere*, the Ministry of Agriculture and Fisheries' inshore research vessel—a side trawler built in the early 1940s—and have met up with *Pukaki* at North Cape to receive packhorse from the pots. This is our opportunity to increase our number of tagged lobsters in the water.

Lying 6 kilometres offshore, we can just make out the diminutive, automated North Cape light on the crest of Murimotu Island. The *Pukaki* crew grapples each buoy, draws it up over the deck railings near the bow, and then, in columns of four, hand-hauls the pot. The last part is the hardest, the pot not sliding up over the rails nearly as easily as the rope. In addition to its inconvenient shape, many

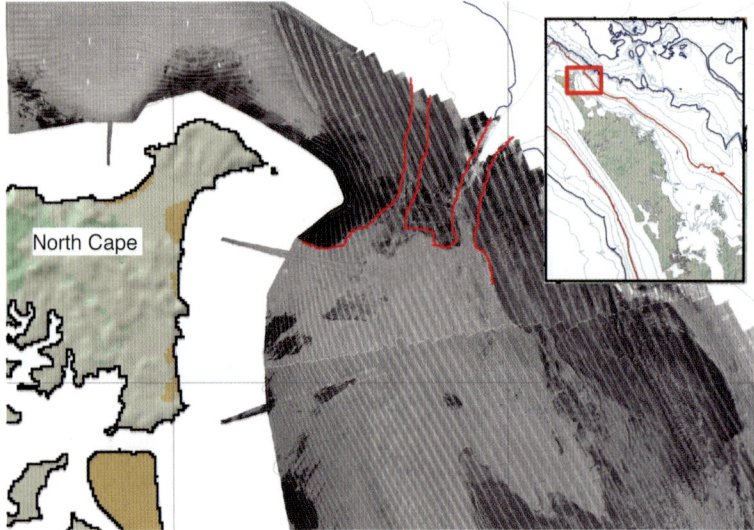

A catch of close to 2 tonnes of mainly small packhorse taken over 2 days late in 1963 on the North Cape Shortie Patch by Cec Ruffell (the first to fish the grounds) on FV *Florence* (left). The hold too is full. Backscatter imaging from echo sounders, which depicts rock and consolidated seafloor more darkly than soft sediments, suggests two alternative channels (outlined in red) that packhorse may take as they round North Cape—if they are not to clamber over rocky bottoms (right).

of the steel-framed pots contain dozens of undersized packhorse, all a consistent 1–1.5 kilograms. Now it's time to tranship the lobsters. Skipper Ken Turner eases *Ikatere* towards the amidships of the naval vessel's port side. He must close the distance between his bow and the side of the *Pukaki* enough for the wicker basket of packhorse to be transferred by hand, but not so close that he scrapes the side of the naval vessel in this moderate swell—tricky with his wheelhouse so far back and the two vessels not moving entirely in concert.

We watch nervously as each time Ken approaches the *Pukaki*, receives a large basket of lobsters and returns the empty one, and then backs off until the next basket is ready to be transferred.

Now it's our job to tag the packhorse lobsters. If the tag is not to be lost at the next moult, it has to be inserted into the flesh: all hard parts are jettisoned when a lobster moults. With smaller lobsters you can, with experience, tag single-handedly, but these packhorse—small as far as they go—are much too large and powerful. You need someone to forcibly draw the tail out from the body a little, at the same time using sheer strength to stifle the tail flapping. Then you must bend the tail to one side. This is because you can't inject the tag into the top centre of the tail for fear of piercing the food canal, an artery, or even the ventral nerve line. Entry from above, but to one side, allows the tag to be inserted into a good hunk of muscle, but it too is not without risk: too deep and the body cavity is penetrated (with unclear, but probably unhappy, complications for the lobster), too shallow and the tag may soon be cast. Sometimes the lobster resists so energetically that the flexible but sturdy membrane between the body and tail tears a little.

Ikatere's Errol Willis and I work our way through the lobsters, which are stored cool, damp and away from the sun under wet sacks. Wearing sturdy plastic gloves, a brick red one on his left hand and for some reason a strange luminous green one on his right, he firmly grips a lobster by its underside as it attempts to flap its tail and skew its sharp pleural spines into his hand. I measure the body with vernier callipers, mumbling '162.2' as I record the length on a sheet of waterproof paper.

'Missing appendages?' I ask.

'Right second leg half missing,' replies Errol.

'Sex?'

'No thanks.'

 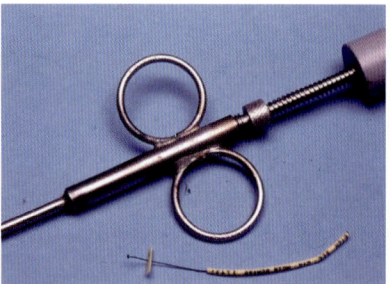

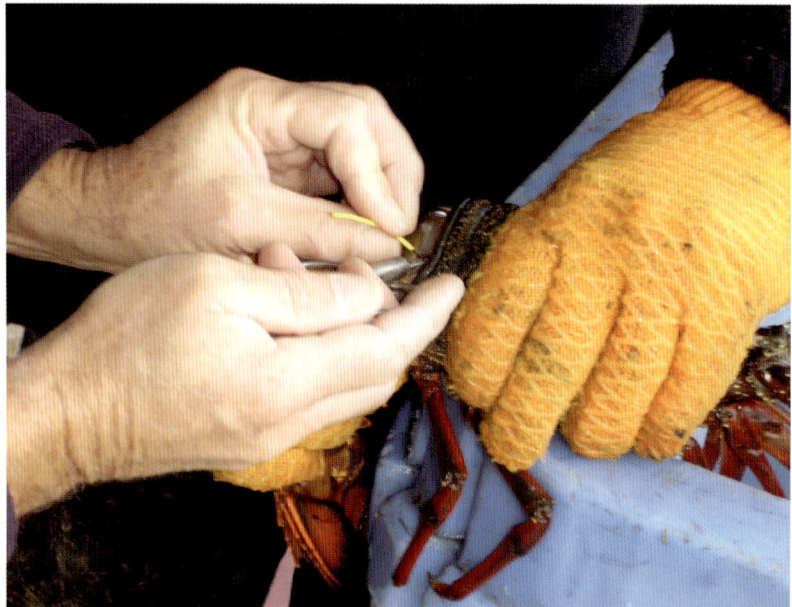

Packhorse were tagged with the western rock lobster tag (centre), the yellow 'spaghetti' section of which extends out from the body (top row). This tag was retained for up to 6 years or more. T-bar anchor tags are much more widely used these days (bottom row). They resemble clothing tags, and self-feed into the gun.

It's male. Errol twists the tail my way, I insert the tag, and the pack is back where it came from.

The focus of the first investigations into the movements of packhorse, in the late 1970s, had been the Shortie Patch off North Cape. Although packhorse a little under and a little above the legal size limit feature strongly—together with much larger lobsters—in pot catches to the west, in the main Far North fishing grounds near Cape Reinga, the Shortie Patch is distinguished by its scarcity of large lobsters. Might most packhorse emigrate west before reaching legal size? Tagging confirmed this to be so. Almost all recaptures of lobsters tagged at North Cape were either at North Cape itself or up to 40 kilometres or so to the west, near Cape Reinga.[4,5] Because it could be argued that the Shortie Patch was a packhorse nursery, and because few red rock lobsters or commercial finfish were being taken there, the then Ministry of Agriculture and Fisheries closed the area to all fishing in October 1977.[4]

The next conundrum was whether or not the generally sparse populations of small packhorse to be found along the east coast south of North Cape were in any way linked to the Shortie Patch, and so in turn to the main fishery grounds. This was a colossal leap in scale, from a mere dozen-kilometre romp to a potentially thousand-kilometre hikoi (march). Our hypothesis was simple: that the entire east coast of the North Island was a nursery for the Far North packhorse grounds.

Most packhorse along the east coast of the North Island are undersized, and the well over 2000 of them tagged in the late 1970s and early 1980s came from pretty much the species' entire main distribution. Not only does the proportion of undersized individuals in pot catches increase with distance south, the undersized packhorse also tend to be smaller on average.[5] And, as at North Cape, they appear often to accumulate in greatest densities south of headlands that extend into the generally south-flowing coastal current. Bream Bay—the island and pinnacle and ocean vista that greets you as you emerge driving north from the twists and turns and bush of the Brynderwyns just south of Whangarei— is one such place where packhorse were tagged. Next, moving south, were the Iles d'Haussez—the Mercury Islands—whose nearby rock outcrops were once *the* hotspot for packhorse catches. Then there was Matakaoa Point, looking over Hicks Bay and one of the easternmost points of mainland New Zealand. Packhorse are still reasonably commonly taken south of East Cape, right down to

Distinguishing the sexes

In male packhorse, the genital pores (Gp) are at the base of the last pair of walking legs (upper left), whereas in females they are at the base of the third pair (upper middle). Also, females have small pincers on the ends of their last pair of walking legs (upper right) which they use to groom their eggs. The swimmerets on the underside of the male tail are made up of one part (lower left), rather than the two of females (lower middle). Once the female is mature the extra branch becomes fringed with setae (hairs) to which the eggs attach (lower right). The sex of all spiny lobsters is differentiated in this manner.

Cook Strait, where catching them is sometimes linked to misfortune. It's said that fishers catching them there won't let the lobster touch their boat, sometimes even cutting off the pot or simply going home—although the precise reasons have never been made clear to me. The tagging sites to represent this stretch of coast were Gisborne and, a little further to the south, the diamond-shaped protrusion that is Mahia Peninsula.

We got back about a tenth of the tagged packhorse, and these lobsters were certainly migratory. Most had moved away from their tagging site by more than 5 kilometres, many a lot further—more than 200 kilometres.[5] With rare exceptions they had travelled northwest, against the prevailing current and towards the Far North. Importantly, there was no evidence for any return migration. Clearly the Far North—the stretch of coast from Tom Bowling Bay in the east to Cape Maria van Diemen to the west, and centred more or less on Cape Reinga—was the primary destination. Although most of the females were immature when tagged—being without setae, the hairs on the swimmerets under the tail to which the eggs attach—most recaptured west of North Cape had by then developed setae, suggesting they had matured. Most lobsters of both sexes recaptured west of North Cape had reached legal size and changed from the dark green of the juvenile to the olive green of the young adult.

The entire east coast of the North Island is the nursery for New Zealand packhorse. Not surprisingly then, divers have reported very small packhorse most often along these shores too, particularly near East Cape, and all packhorse larvae and postlarvae have come from the east of the North Island. In contrast, the only place where significant quantities of large mature packhorse have been taken in the past few decades is the Far North. Although notable numbers were taken along the east coast as far south as White Island in the early days of the commercial fishery, these populations were quickly fished down and essentially extinguished ages ago.

But the extent of these movements towards the Far North was modest compared with some. A small proportion of the tagged packhorse moved west from the Shortie Patch across the northern tip of the North Island, then continued south. The greatest straight-line distance covered by an individual in its march was an astonishing 1070 kilometres.[5] Tagged at North Cape and recaptured near Knights Point in south Westland four and a half years later, this packhorse had walked south at the very least half a kilometre each day. Another seven (including those two near New Plymouth) were recaptured at points between Cape Maria van Diemen and Knights Point. Probably then, many if not most of the packhorse to be found along the west coast of New Zealand arrived there after migrating from the north, rather than being derived from local postlarval settlement. So, individuals found from time to time in Foveaux Strait are likely to have migrated between 1400 and 2400 kilometres, depending on just where on the east coast of the North Island they had settled out of the plankton. However, for most the destination is the Far North, an average straight-line migration of more than 500 kilometres—the distance between Cape Reinga and a point halfway down the east coast to Cook Strait.

Large numbers of packhorse lobsters gallivanting great distances along the open seafloor under the full scrutiny of predators? Nature is not so asinine as to put any of its own through such an ordeal without good reason. The migration along the east coast to the main breeding grounds of the Far North can be viewed as contranatant—and essential to the persistence of the species in New Zealand waters. The adjective 'contranatant' was coined by Alexander Meek in 1915 to describe the journey of juvenile fish back to their breeding areas after being carried away by ocean currents.[6] For packhorse this journey is a walk along the seafloor rather than a swim in the water column. It counters the southward and eastward drift of the poor-swimming and long-lived larval phase. Unless at least some lobsters return upstream toward their place of hatching to counter this displacement, packhorse could eventually die out in New Zealand waters. (The walk south along the west coast of the country by a small proportion of the packhorse is presumably triggered by the weak northward net flow that exists along much of this shore.)

This great propensity to migrate alongshore as they approach maturity means that packhorse are almost certainly the most migratory spiny lobsters on Earth—and may well be the most peripatetic of all marine invertebrates. This is in terms of both the proportions migrating and the distances covered. A similar migration is seen in juvenile *Sagmariasus verreauxi* in New South Wales, the only other mainland place where packhorse are found—and where they are known as eastern rock lobsters.[7] Here, the distance between the main breeding areas and the main settlement areas is also about 500 kilometres.

Wide Wanderings   35

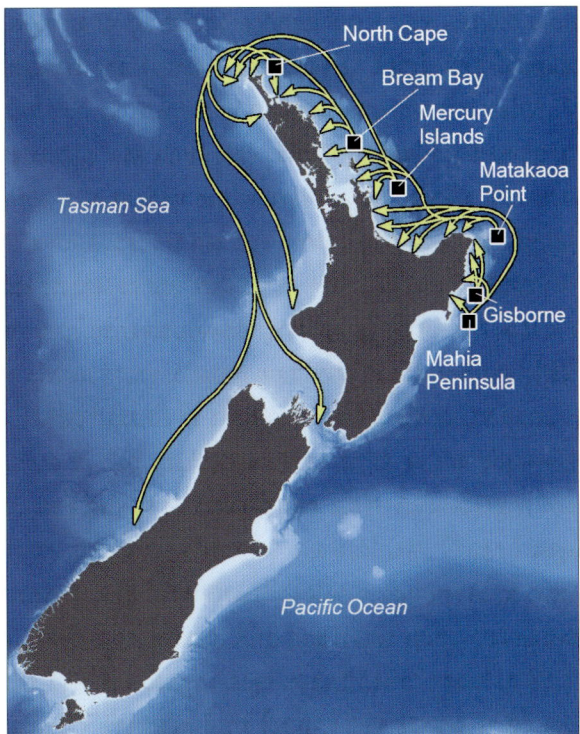

Movements of tagged undersized and maturing packhorse. Boxes show tagging sites. Note that the Cape Reinga area in the Far North is the primary destination, few packhorse migrating south along the west coast.

Of the 55 species of spiny lobsters on Earth, only a handful of others are known to migrate alongshore to anything like the same extent as packhorse. And one of these is a close relative, the red rock lobster *Jasus edwardsii*, with which packhorse once shared its generic name. This species is found throughout the coasts of New Zealand and southern Australia. (The long-distance migrations of four other spiny lobster species are illustrated towards the end of Chapter 4, the movements being crucial to their respective recruitment mechanisms.)

In southern New Zealand, during spring and early summer, some of the red rock lobsters approaching maturity move various distances against the general direction of the coastal current. This takes some through to the west coast of the South Island. Discovered by marine biologist Bob Street in the 1960s, the migration is against the flow of the Southland Current: south along the east coasts of the South Island and Stewart Island; north from the Snares Islands and the west coast of Stewart Island; and generally north towards Fiordland and south Westland.[8] The greatest distance covered by a tagged individual was almost 500 kilometres; the fastest ones marched 7 kilometres each day (although mean rates of movement were much lower). The likely explanation for these migrations is that, exactly as for packhorse, a proportion of the juveniles move to counter the drift of the larvae. But in contrast to packhorse, the red rock lobsters in southern New Zealand appear to have no single, major destination, and although the proportion migrating varies a lot from year to year, it is generally small. Tagged lobsters moved various distances south from points between Banks Peninsula and Stewart Island, and north from points between the Snares Islands and Fiordland. No individual tagged in Otago ever turned up in Fiordland, and few crossed Foveaux Strait from Stewart Island. This pattern of migration leads to the remarkable possibility of red rock lobsters moving north along the southwest coast of the South Island running into a packhorse plodding its way south.

Mass alongshore migrations, all in the same general direction, don't appear to take place among red rock lobsters in other parts of New Zealand[9]—but

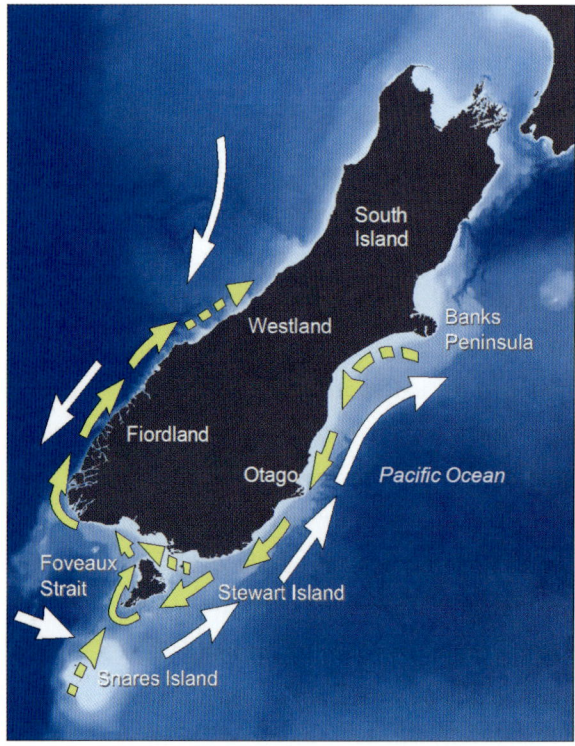

Red rock lobster migrations (yellow arrows) in southern New Zealand, against the general coastal flow (the Southland Current, white arrows), led to some large trawl catches in south Westland in the early days.

this remains unconfirmed because too few immature lobsters have been tagged. Except in the southeast of New Zealand, the size at onset of breeding is so small that the immature lobsters are difficult to catch and are too small to cope very well with being tagged.

Bob Street (left) discovered the migration of red rock lobsters in southern New Zealand, which in places takes the form of queues.

Packhorse approaching maturity migrate in a highly directional and purposeful manner against the general flow of the coastal current, over several months, towards the Far North of New Zealand, and red rock lobsters in the south move similarly south, then west, then north—even though for both their exact routes hour-by-hour are quite unknown. What is their lodestar? One could imagine the lobsters detecting the current as wind on the face. But coastal currents are seldom, if ever, straight-line flows. Instead there is eddying and reversal according to the shoreline shape, the prevailing weather, and the offshore oceanography. Andrew Jeffs of the University of Auckland has long been interested in the navigational cues used by spiny lobsters, and here is how he sees things, with particular reference to packhorse:

> Imagine being taken from home blindfolded to a location over 35 kilometres away that is completely unfamiliar, and then being asked to walk home. Your chances of getting the correct direction are very low, unlike spiny lobsters. This is because spiny lobsters are one of a small and elite group of animals that possess true navigational skills—the ability to orient to a

known location from an unfamiliar place without retracing clues picked up during the outward journey.[10] How lobsters navigate remains somewhat of a mystery. However, experiments with magnets suggest lobsters have an ability to use subtle differences in the Earth's magnetic field to figure out their direction at otherwise unfamiliar locations.[11] If packhorse possess a magnetic compass, migrating north would be an easy task.

But there are other possible cues. The bodies of lobsters—especially the antennules and antennae—are covered in thousands of chemical sensory hairs. These hairs are capable of detecting minute amounts of dissolved chemicals in seawater and are vitally important in sniffing out food from a long distance, such as bait in a distant lobster pot. The same hairs can be used for sensing changes in water chemistry and smells that may also help to set a migrational course. For packhorse lobster migrating north this chemical sense may detect the specific chemical signature of water that originates from the north.

In coastal waters where most spiny lobsters dwell, waves tend to move toward the coast. This creates ridges that run parallel to the coast along any seafloor made of mud or sand—an obvious signpost to follow. Other underwater features or contours such as canyons, ridges or shelves may also provide a pathway for lobsters to follow over a longer distance.

Regardless of these obvious signposts, lobsters are especially sensitive to water movements, such as wave motion and water currents. Detecting these water movements may also help lobsters to steer underwater. For example, the migration paths of several species of spiny lobster tend to follow, or run directly against, the prevailing ocean current. The south-flowing East Auckland Current along the east coast of the North Island may provide a useful directional cue for migrating lobsters. Even though this current is far from being steady and uni-directional, the lobsters may still use it to orientate, simply taking longer to reach their upstream destination than if they went directly.

Underwater sounds can also provide directional cues over quite long distances because certain sounds travel a very long way in water. Waves crashing on a shoreline can be heard kilometres offshore, as can the noisy nocturnal animals that live on rocky reefs. There is evidence that various species of crustacean can detect these underwater sounds.[12]

Packhorse may use any one or several of these directional cues to navigate back to their breeding grounds. For a brief part of their lives they succumb to an innate desire to migrate alongshore, with remarkable directional sense.

Moving great distances has so far been the prerogative of the adolescent spiny lobsters. Some potent switch is turned on just before the lobsters mature, and then is turned off soon after—without revealing, so far, very much of the mechanism involved. This is not to say though that once they reach the breeding grounds the young adults are lazy. Their much more local ambulations can include seasonal return migrations associated with moulting, reproduction, and even feeding; movements initiated by change in the ocean itself brought about by such things as variations in atmospheric pressure; foraging excursions from home dens; and simple random wandering. But the business of sorting out each of the different types of movement—and distinguishing them from the contranatant migrations—is not necessarily straightforward. The movement of an individual from one point to another might be for one or several of these reasons—with a dose of observation error to boot.

The seasonal movements of packhorse, once on the Far North grounds, are tantalisingly enigmatic. Movements to do with moulting and reproducing take place over a scale of kilometres, but because the lobsters come and go from the same spots year after year, they are difficult to pin down. Commercial fishing suggests that both small and large packhorse have accumulated in shallow (less than 60 metres deep) inshore waters of the Far North by October.[5] The mature females among them are bearing eggs and so must have recently moulted—a recent moult ensures that the setae are long and intact for attachment of the new clutch of eggs. During December and January the lobsters move out to deeper waters, to depths of at least 100 metres, where the eggs hatch. These waters are more tidal, which presumably enhances larval dispersal.

The trail then becomes very hazy. On the one hand, I was told that both small and large packhorse appear in pots set in deep waters increasingly further to the west, off Capes Reinga and Maria van Diemen, as the season progresses. Eventually they reach waters perhaps 200 metres deep, where by May they no longer enter pots. The tag returns support this, as does the informal tagging of fisherman Bill Hopkins. (He touched up packhorse in Spirits and Tom Bowling bays with black paint and later in the same season some of them turned up off Cape Reinga.) Then, for these lobsters to be seen again inshore the following season, they must make their way back east between May and October.

On the other hand, probably not all lobsters move west, some—both large and small—simply moving into nearby deeper waters at egg-hatching time. Either way, packhorse in the Far North are remarkable in that each year many—probably most—routinely traverse a depth spread of as much as 200 metres. This journey takes them over a variety of seafloors, from inshore sands and reef to fissured hard bottom to open ground best described as sand interspersed with areas of low reef and boulders. During this time they must avoid all manner of predators—including humans.

Many, if not most, species of spiny lobster worldwide move seasonally inshore and off as they moult and reproduce, in a manner not dissimilar to packhorse in the Far North of New Zealand. Female red rock lobsters move inshore during autumn to moult, mating and laying eggs soon after; the males move inshore to moult in spring.[13] Moulting inshore may be to make use of water movement in the shallows to assist in the casting of the old shell. In spring, towards the end of egg-bearing, females with eggs often move to deep, seaward edges of reefs and near headlands where strong water movements presumably enhance larval dispersal.[9]

One of the easier movements to pin down is the non-return migration of small juveniles from inshore nurseries where settlement has taken place to deeper offshore waters where the breeders live. It is most stark where the continental shelf is wide. In Western Australia, 3–4-year-old juvenile western rock lobsters *Panulirus cygnus* move en masse 30–50 kilometres, eagerly pursued by fishers because many of the lobsters have reached legal size by then.[7] The mass migrations of the Caribbean spiny lobster *Panulirus argus* into deeper waters in response to approaching storms are often to be seen in wildlife documentaries. They take the form of single-file migrations on a brightly white seafloor palette, the lobsters from time to time rounding up to form defensive pods, their flailing antennae warding off predators.

A remarkable short—and certainly final—migration takes place from time to time among Cape rock lobster *Jasus lalandii* on the southwest coast of southern Africa.[14] Decomposing phytoplankton blooms, made up of great concentrations of microscopic algae, regularly drag down the nearshore dissolved oxygen levels in the already often suffocating Benguela Current waters. The lobsters move into the shallows in their bid for oxygen, only to strand and succumb.

Poorly oxygenated water brought about by decomposing phytoplankton led to the stranding of 1500 tonnes of Cape rock lobsters at Elands Bay, South Africa in 1997.[15] Regular strandings of hundreds of tonnes need to be taken into account in stock assessments.

CHAPTER 2

SO, JUST WHAT ARE PACKHORSE?

A drawing is often better than a photograph for illustrating the features of an animal, or its parts. It's not that the illustrator seeks to mislead, but more that the relevant features can be highlighted by subtle use of stippling and shading. Biologist Richard Pike produced the definitive illustration of the packhorse spiny lobster in 1969.[1] This drawing (and his one of the red rock lobster *Jasus edwardsii*) was chosen by renowned Dutch crustacean taxonomist Lipke Holthuis for his 1991 *Marine Lobsters of the World*—a handbook indispensable for all students of the world of spiny lobsters.[2] And, with the lobster's bulging middle giving it the appearance of a beast of burden, the drawing is useful in illustrating the grounds for the common name 'packhorse'.

Illustrations such as these are most often based on preserved specimens housed in organised collections. There, new discoveries are deposited, a representative specimen called the 'holotype' sooner or later being formally described and illustrated, and the account published. In the Museum of New Zealand Te Papa Tongarewa natural history collections, the slightly musty smell is augmented by the background odour of preservative. On giant stacks of shelves—so heavy they run on their own tracks, with geared mechanisms to allow them to be moved at all—there is glass jar after glass bottle of ethanol-preserved marine animal. Some of the containers are large—like those used to preserve in-season produce—but others only jam-jar size. The smallest biological specimens are in glass vials, held within not by a lid or cork but by a stuffing of cotton wool; these tiny containers in turn are held within larger, sealed jars of preservative. The 20-litre white plastic pails—larger versions of those used for paint—seem far less interesting because there is not even the shadow of their contents to be seen until the lid has been prised off and the material brought to the surface. All of the specimens have a label, the name of the collector, the date and the place carefully inscribed in black on thick, durable paper. The older ones, often elegantly hand-written in Indian ink, provide a satisfying link with another time and place and are as collectable as a celebrity's signature. More recent labels are anonymously computer-generated.

Spiny lobsters, like other animals that are typically preserved in fluid, are first fixed in formalin. This solution—more widely known as embalming fluid—suspends biochemical reactions, halts bacterial activity, and increases the stability and strength of tissues. Larger specimens—if their tissues are not to be removed—must be injected with formalin so that their core is fixed before it rots. Then, after a few days or weeks, the lobster is drained and rinsed, to be stored in ethanol—a solution much kinder to hand and lung. This technique preserves the form of hard-bodied animals well, but seldom their colour, which rapidly dulls.

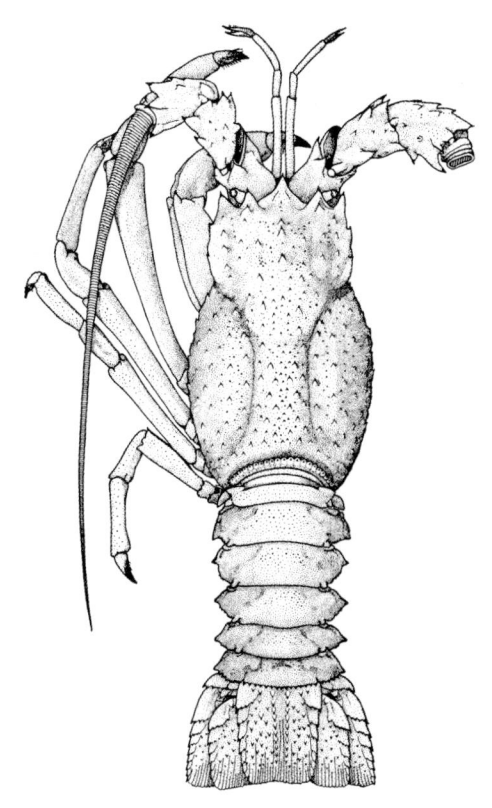

Richard Pike's 1969 drawing of a large male *Sagmariasus verreauxi*[1] dramatically illustrates provenance for the common name 'packhorse'.

Animals with durable outsides—such as shellfish—may be filed away dry (but this means that their soft tissues are not available for study). Some of the spiny lobsters in Te Papa have been shelved in this manner.

And one dried-out carcass, in a large, acid-free cardboard box crafted specially for it, is of immense significance to packhorse.

The packhorse spiny lobster—also known in New Zealand as a pack, packhorse, packie, packhorse rock lobster, packhorse crayfish, green, pawharu or koura—is a species of spiny (alternatively, 'rock') lobster, and so belongs in the

In juvenile packhorse the branchial regions on the sides of the carapace are far less developed than they are in the adult, and the lobsters are distinctly green.

family Palinuridae. Spiny lobsters are closely related to, but very different from, the clawed marine lobsters of North America and Europe. Most noticeably they lack large pincers, instead possessing simply oversized front legs. Strictly speaking, the term 'crayfish' is reserved for the clawed, freshwater lobsters.

Spiny lobsters live all round the world, from the Equator north and south to 55° or so, at depths ranging from just under the low water mark—at night even in the intertidal zone—to well below 1000 metres.[2] They live in a diverse variety of habitats—rock, sand, mud, weed and coral—and have evolved a broad and almost perplexingly dramatic palette of colours and patterns to allow them to blend in with their backgrounds. Most exist in shallow waters associated with large land masses. But a few species are widespread at great depths on both trawlable and rough seafloor, and several species live—seemingly precariously—around small remote islands and seamounts. A handful so far have even been found confined entirely to seamounts and ridges well away from any emergent land mass.

The spiny lobster family (Palinuridae) is one of more than 150 families in the order Decapoda, decapods being 10-legged—or, more precisely, 10-footed—crustaceans. The order also takes in the shrimps, prawns and crabs. Their bodies consist of two main parts, the cephalothorax and the abdomen or tail. Spiny lobsters are distinguished from their closest relatives, the slipper lobsters (family Scyllaridae), by their feelers, which are long, and whip- or spear-like, rather than flat and plate-like.

The precise hierarchical taxonomic placement of packhorse follows, bold font being used only to make the detail clearer. The names and dates associated with each level—and why some are enclosed in parentheses and others not—are explained shortly.

Order **Decapoda** Latreille, 1802
Suborder **Pleocyemata** Burkenroad, 1963
Infraorder **Achelata** Scholtz & Richter, 1995
Family **Palinuridae** Latreille, 1802
Genus ***Sagmariasus*** Holthuis, 1991
Genus and species ***Sagmariasus verreauxi*** (H. Milne Edwards, 1851)

Taxonomy—the practice and science of classifying living things—is of necessity a very precise discipline, with agreed rules that govern name format and precedence. Now more than ever do we need to know categorically what species we are dealing with. Fishery managers these days are concerned with the resilience of populations of individual species, not groups of species. This is particularly because similar-looking and even closely related species can have very different life history features such as how quickly they grow and how long they live. Marine products on the international market need to be labelled correctly. And, perhaps most importantly, jurisdictions that are signatories to the United Nations Convention on Biological Diversity commit to conserving marine biodiversity within their exclusive economic zones and revealing the risks the creatures face—particularly those risks brought about by humans. This means knowing Fanny from Fortescue.

Rick Webber's job at Te Papa is to curate the crustaceans and here he explains how scientific names are conceived.

> We think of packhorse spiny lobsters as just that. It's perfectly clear which species we are referring to, so why the scientific name? *Sagmariasus verreauxi* is also a bit of a mouthful, the spelling makes it difficult to pronounce or remember, and in formal publications it even has the name of the person who named it, and when. But, given all that, the 'Latin name' *Sagmariasus verreauxi* follows a naming system invented 250 years ago which has not changed greatly to the present day, so there has to be something in it. And in fact, there is a group called the International Commission on Zoological Nomenclature which publishes and updates rules on how and how not to apply animal names. The rules are listed in the International Code of Zoological Nomenclature (found on the web under its acronym ICZN).
>
> It was Carl Linnaeus who, in the mid 1700s, introduced this system of binomial nomenclature for naming animals and plants. Scientific names weren't new in Linnaeus's day; other great minds were grappling with the problem of how to create an organised system for naming the rapidly

increasing number of species being collected world wide in the 'age of discovery'. Various naming systems were invented but they did not have the simplicity or scope of Linnaeus's approach.

Sagmariasus is the genus name, *verreauxi* is the species name, the name in brackets is of the Frenchman Henri Milne Edwards who gave it the name *verreauxi,* and the date that follows is the year he published this name. The genus always goes before the species name and always begins with a capital; the species name these days always begins in lower case (although it used to start in upper case if named after a person or place). In the case of *S. verreauxi*, the author's name and date are in brackets to indicate that the genus name has changed from the one it was originally placed in by Milne Edwards. Its first name was *Palinurus Verreauxi* H. Milne Edwards, 1851. It joined a number of other species in *Palinurus* to which Milne Edwards felt *verreauxi* was closely related. The packhorse was subsequently removed from *Palinurus* and placed in the genus *Jasus*. Thus it became *Jasus verreauxi* (H. Milne Edwards, 1851) with brackets to indicate the change. More recently it was moved again and given its own genus, *Sagmariasus*, with the brackets, of course, retained.

It would be nice to think that giving a name to a species is all there is to it but already it is obvious that names change. It should also be evident that the grouping of species in one or other genus, or family or order, reflects apparent evolutionary relationships between species and species groups.

Of course, there once were far fewer named species than there are now, and they were identified, and compared and contrasted with other species, on the basis of their morphology. These physical characters, such as the presence and positions of spines or the lengths of limbs etcetera, are still regarded as essential to taxonomy, but other tools for distinguishing species and species groups have developed along the way. Nowadays molecular analysis, DNA, the stages of development, even behaviour and sperm morphology, are put to use. The result is that identifications become ever more precise which often means that what used to be accepted as a well-defined species turns out to be more than one species, or alternatively, is the same as another species. *S. verreauxi* has belonged to two different genera in the past. Over time its distinctive characteristics, and those of others in the same genus, became better defined so it and the others were shifted into a more appropriate genus.

The dynamic outcomes of continuous taxonomic investigation do not stop with species and genera. There is a hierarchy of ever larger and more inclusive groups above *Sagmariasus verreauxi*, from the family Palinuridae (spiny lobsters) to infraorder Achelata (spiny lobsters and slipper lobsters) to the order Decapoda (shrimps, prawns, crabs and lobsters) and so on up to the subphylum Crustacea. None of these groups is static and all are the subject of research and often fierce debate. Very academic people are constantly engaged in redefining and reorganising this hierarchy into something that better represents the evolutionary relationships between the species and groups at all levels. The name Achelata (meaning 'without chelae, or pincers') for instance, was only set up in 1995 after a very detailed study of many comparable morphological features indicated that spiny lobsters and slipper lobsters are closely related and closer to each other than to the polychelids, lobsters which used to be placed with them in an older infraorder called Palinura.

The packhorse has therefore, like most animal species, undergone many changes in scientific name over the decades as researchers tried to establish its proper taxonomic position. This has taken place with a level of confusion, ambiguity and mystery—all of it crying out for closer scrutiny.

The Treaty of Waitangi had long been signed when the packhorse spiny lobster was first recorded scientifically. Henri Milne Edwards was from France, his rather un-French last names coming about through his father having been English. It is said that Henri was his father's 27th child, and to avoid confusion with his prodigiously numerous relatives, he added his middle name Milne to the surname Edwards.[3] (He usually wrote 'Milne Edwards', without a hyphen, but his son Alphonse Milne-Edwards, who also described spiny lobsters, always used a hyphen.) Henri did not visit either New Zealand or Australia. Rather, we presume he examined and named this new lobster *Palinurus Verreauxi* from the collections of the Muséum national d'Histoire naturelle in Paris, where it is believed the holotype was once housed. Frustratingly he provided no information on the origin of the lobster. His very brief paper refers to 'Langouste de Verreaux', the lobster of Verreaux—presumably for Jules Pierre Verreaux, a professional collector of, and trader in, natural history specimens, which was all the rage in those days.[4] (Jules went on to become an assistant naturalist at the Paris museum.) Henri limited his description to the front part of the body, which clearly aligned it with other spiny lobsters that were at the time known as *Palinurus*.[5]

In 1862, the Austrian marine biologist Camil Heller referred to packhorse as a new spiny lobster altogether, *Palinurus Hügelii*.[6] And the name had nothing to do with size. The lobster had been collected, probably from near Sydney, by a certain Karl Alexander Anselm Freiherr von Hügel, baron of the German Empire. Not until much later was it was realised that both he and Henri were talking about the same animal. But oddly, Camil reported that the species had been collected in the Indian Ocean—twaddle because neither this species nor any like it is found there. The likely source of the locational error is that Karl, in addition to visiting both New Zealand and Australia, spent time in India and in other places bordering the Indian Ocean.[2]

Confusion continued. Thomas Kirk, botanist and then Assistant in the Colonial Museum in Wellington, New Zealand, in 1880 described a lobster obtained by Mr J. Buchanan in December 1877 at 'Whaingaroa, a small harbour on the West Coast of the North Island'.[7] It had been lodged in the museum bearing the name '*Palinurus hügelii* var. *tumidus*'. Thomas noted that the specimen did indeed look very similar to *P. hügelii* 'of the Indian Ocean' but felt that it had characters sufficiently distinct to justify its elevation to the rank of species. He particularly remarked on its size and how *giganteus* might have been a more appropriate name: it was more than 60 centimetres from between the feelers to the tip of the tail, and the very swollen carapace was over half a metre in circumference. He called it *Palinurus tumidus*, this specimen being the holotype for the species.

Which brings us back to that large cardboard box in Te Papa's natural history collection. It's 130 years on, yet the specimen is still in reasonable shape. Trussed with white fabric tape tied in bows, it has been varnished at some point to help conserve it. Still, there is sign of decay: tell-tale patches of brown dust show that the museum beetle—a type of carpet beetle that can play havoc with dried animal collections—has been at work some time in the past. The label, browned over time, is shown to the right of this sentence. Note that the letter *I* has been added by hand between the *A* and *N* of WHANGAROA.

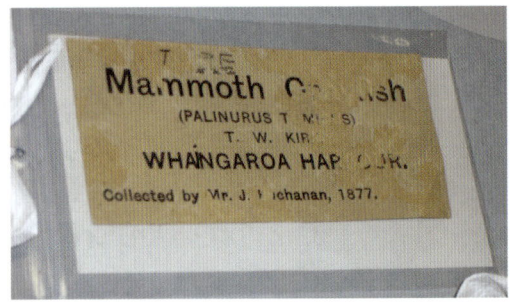

Contained within the label pocket is a more recent label, handwritten.

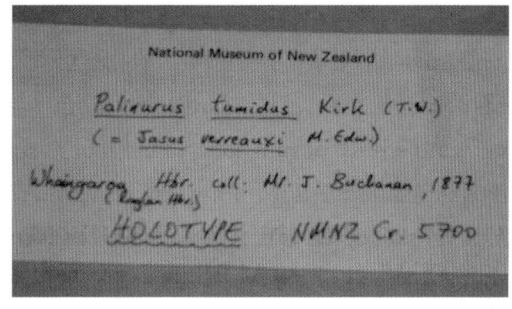

There is no doubt that the specimen is a large packhorse, but the location of the collection is curious. One presumes that the change in name from 'Whangaroa' to 'Whaingaroa' was made early on—before Thomas Kirk published his description. But Whaingaroa, an early name for what is now more widely known as Raglan Harbour, on the west coast of the North Island, is a much less likely origin for this packhorse than Whangaroa Harbour. Whangaroa Harbour is on the east coast of Northland and completely within the common range of large packhorse. Furthermore, zoologist Gilbert Archey later wrote that the specimen had been 'obtained from Waingaroa [sic], in the north of Auckland'.[8]

A clue to the actual source location of this lobster should lie with its collector. John Buchanan was first and foremost a botanist—several plants are named after him—but he also contributed to the fields of geology and zoology.[9] He certainly did spend time in the Raglan area, but during the summer of 1865–66 he, together with Thomas Kirk, explored the botany of the lands between Whangarei and North Cape. This took in Whangaroa. The exact origin of NMNZ Cr 5700 may, therefore, never be known for certain, but my money is with Northland's Whangaroa.

Soon after, in 1882, Scottish-born biologist William Haswell, who spent most of his career at universities and museums in the east of Australia (and quite a lot of time at the Sydney fish markets), pointed out that there was no real difference between *P. tumidus* and *P. hügelii*. The main difference was the shorter tail tip of *Palinurus tumidus*, which he supposed was due to wear or other damage, or simply 'an artist's slip'.[10]

Among the more important of the several name changes that followed was the 1883 placement of packhorse in the new genus *Jasus* by T. Jeffery Parker, who was Professor of Biology at the University of Otago, Dunedin.[11] On his way out to New Zealand he had stopped off at the Cape of Good Hope where local spiny lobsters (then known as *Palinurus lalandii*) were brought aboard. He noticed there was no stridulating organ (a sound-producing structure at the base of the feeler) on these lobsters—which otherwise looked very similar to the common European species of *Palinurus*. There was therefore a clear distinction between spiny lobsters with a stridulating organ and those without one: the stridulators (Stridentes) were *Palinurus* and *Panulirus* and the non-stridulators (Silentes) *Jasus*. This necessitated change to the generic name for packhorse—which of course lacks a stridulating organ—but the origin of the name *Jasus* seems not to have been provided by T. Jeffery. Lipke Holthuis suggests it may refer to *Iasus*, the Latin name of a locality in southwest Turkey;[2] perhaps T. Jeffery once had a holiday home there.

Among Rick Webber's responsibilities at Te Papa is curating this 130-year old museum specimen, recorded as the holotype of *Palinurus tumidus*—but in fact the packhorse *Sagmariasus verreauxi*. (The feelers have been placed beneath the lobster.)

Much later, in 1991, Lipke recognised the need to distinguish *Jasus verreauxi* from the other *Jasus* species because it was morphologically so different. He erected two subgenera, *Jasus* (*Sagmariasus*) for *J. verreauxi* and *Jasus* (*Jasus*) for all other species.[2] The name *Sagmariasus* came from a clever combination of the Greek *sagmarion*, meaning packhorse, and *Jasus*. Together they acknowledged the resemblance of large individuals to a laden horse and the closeness of the subgenus to *Jasus*.

 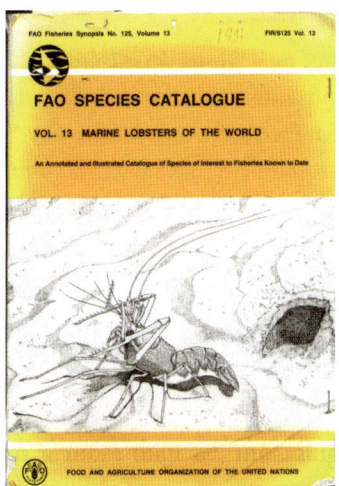

Renowned Dutch crustacean taxonomist Lipke Holthius and his catalogue of the marine lobsters of the world.

In 2001, *Sagmariasus* was elevated to full generic status, and the packhorse spiny lobster became *Sagmariasus verreauxi* (H. Milne Edwards, 1851).[12] *S. verreauxi* was not just morphologically distinct from *Jasus* lobsters; it also differed greatly in numerous other ways, including genetically.[13] But this may not be the end to the matter. A provisional result has suggested slight genetic differences between *S. verreauxi* stocks in New Zealand and Australia.[14] If this difference proves real then it means that the two stocks should be considered separate species. Because Henri Milne Edwards's type material apparently came from New South Wales, the Australian species would be *S. verreauxi* and the New Zealand one *S. hügelii* (Heller, 1862) or *S. tumidus* (Kirk, 1880). But more sampling is needed in order to be sure.

As we delve further into the remarkable world of spiny lobsters, we'll meet up again with those cousins of packhorse—the ones with which it once shared a generic name. And in the table that follows we see exactly how both *Sagmariasus* and *Jasus* fit into that world.

φφφ

What it is that inspires authors to give animals their particular names can be gleaned from the list of *Jasus* species.

> Cape rock lobster *Jasus lalandii* (H. Milne Edwards, 1837)
> Foundation Seamounts rock lobster *Jasus caveorum* Webber & Booth, 1995
> Juan Fernández rock lobster *Jasus frontalis* (H. Milne Edwards, 1837)
> Red/southern rock lobster *Jasus edwardsii* (Hutton, 1875)
> St. Paul rock lobster *Jasus paulensis* (Heller, 1862)
> Tristan rock lobster *Jasus tristani* Holthuis, 1963

The specific names *paulensis* and *tristani* are obviously to do with where the lobsters are found, as is *caveorum* (to be explained later). The names *edwardsii* and *lalandii* are for people, the first for Alphonse Milne-Edwards and the latter for Pierre de la Lande, a naturalist and explorer from the same part of the world.[2] Spiny lobster names that highlight anatomical features are particularly useful, because they assist in recognising species and in memorising names. An example is the unicorn blunthorn lobster *Palinustus unicornutus*, with its single spine at the front of the carapace compared with more complex armature in other species of the genus. (The California spiny lobster *Panulirus interruptus* is named for no other reason than that 'the transverse grooves on the tail segments are abruptly and broadly interrupted across the midline'.)

The table below of the world's spiny lobster genera contains a curious source of considerable confusion. When the genus *Palinurus* was split into three genera, two new names that are anagrams of *Palinurus* were used—*Panulirus* and *Linuparus*. *Linuparus* is sufficiently different from the others to cause no difficulties, but *Panulirus* and *Palinurus* are frequently confused.[2] One suspects wickedness here. Also, *Puerulus* as a generic name, erected in 1897, invariably arouses a second take, as will become clear shortly.

Earth's spiny lobster (family Palinuridae) genera.[2] Silentes genera are highlighted in bold (they lack the noise-making stridulating organ at the base of their feelers); the rest belong to the Stridentes. (There is a full list of species in the next chapter.) Note that, in line with recent analyses,[15,16] all clawless lobsters once known as 'coral' lobsters and placed in family Synaxidae (*Palinurellus* and *Palibythus*) are here considered to be spiny lobsters. Shallow is less than 200 metres; Intermediate is 200–400 metres; and Deep is beyond 400 metres (the deepest-dwelling species, the Cape jagged lobster *Projasus parkeri,* can be found down to 3000 metres or more).

Genus	No. of species	Depth	Main seafloor type
Jasus	6	Shallow to intermediate	Reef
Justitia	1	Shallow to intermediate	Reef
Linuparus	3	Shallow to intermediate	Reef, sand, mud
Nupalirus	3	Shallow to intermediate	Reef
Palibythus	1	Intermediate	Reef (probably)
Palinurellus	2	Shallow	Reef
Palinurus	6	Shallow to deep	Reef, mud, sand
Palinustus	5	Shallow to intermediate	Mud, sand, reef
Panulirus	21	Shallow	Reef, sand, mud
Projasus	2	Intermediate to deep	Reef
Puerulus	4	Intermediate to deep	Mud, sand, shell
Sagmariasus	1	Shallow	Reef, sand

φφφ

The packhorse of today is the result of millions of years of evolution, but there is a dearth of preserved material to reveal its sequence. The only fossil we have was discovered by Mr Ulrich of the Golden Bay Cement Works in the Tarakohe Marl Pit in Golden Bay, near the northwest tip of the South Island of New Zealand, half a century ago. It was preserved in its natural position on a slab of fine grey argillaceous sandstone, only part of its front damaged. There is no way of knowing what it had been doing that day, or why it had been out there on fine sediment at all, but it appears to have moulted shortly before it became entombed, as it is incompletely calcified.

There was no doubt in the mind of Martin Glaessner of the University of Adelaide as to the close relationship between this fossil and the packhorse.[17] But there were differences between them: the fossil had fewer spines, and very different grooves on its tail. Small compared with the marketed packhorse of today, it was nevertheless a substantial lobster at 120 millimetres carapace length. And the epoch in which it was going about its business—20 million years ago—was revealed by the Lower Miocene life forms found with it.

The fossil *Sagmariasus flemingi* discovered in a marl pit near Nelson.

Martin Glaessner named this holotype, and sole specimen, *Jasus flemingi*, in honour of the Chief Paleontologist of the New Zealand Geological Survey of the time, Charles Fleming. With the recent change in generic name, this lobster has become *Sagmariasus flemingi* (Glaessner, 1960). It is likely that *S. flemingi* is the forefather of the packhorse of today—probably having emerged at about the same time as, or soon after, the appearance of moa and other flightless birds for which New Zealand is well known.

The animals we call lobsters are reptant (crawling) decapod crustaceans that are thought to have emerged more than 250 MYA (million years ago), during the Permian period—about when the first true seed plants and the early dinosaurs appeared. Their body form and genetics suggest that lobsters are paraphyletic—consisting of individuals descended from more than one common ancestor.[18] And the ones that we know most commonly as lobsters and crayfish can be divided into two monophyletic descendent clades (groups). The first, the Homarida, includes the clawed lobster families—the Enoplometopidae (reef lobsters) and the Nephropidae (clawed lobsters)—with a sister clade the Astacida (freshwater crayfish). The second group of lobsters, the Achelata clade, are the 'clawless' lobsters: the Palinuridae (spiny lobsters) and the Scyllaridae (slipper lobsters). The Achelata emerged during the Late Triassic period, perhaps 200 MYA, early in the reign of the dinosaurs and about when the first mammals appeared.[19]

Spiny lobsters evolved in the face of dramatic change in the shape and character of the oceans. They arose in the early Mesozoic era in the Atlantic-European region of the vast warm Tethys Ocean and had diverged by the Jurassic period into two groups—the Stridentes, with a stridulating organ at the base of each antenna, and the Silentes, without one.[21] The northward drift of the southern continental plates and the break-up of the Tethys Ocean in the early Tertiary era created many new habitats into which the family radiated. At the same time there is strong evidence for a shift from deeper waters to shallower waters with more varying characteristics, leading to the emergence of physical features that were presumably adaptive in avoiding predators in the new, better-lit habitats and in improving aeration of the egg mass. These adaptations included wider and more elevated supraorbital processes (the armour above and protecting the eyes), longer and higher eyestalks, larger pleopods (swimmerets under the tail), and a more rounded and sturdier body.[22] (But, just to complicate things, subsequent retreat into deeper waters may have taken place in such genera as *Linuparus*.)

Western Australian Ray George considers *Projasus* (a species of which, *P. parkeri*, is found in New Zealand waters) to be very similar to the ancestral Silentes stock. The genus originated, perhaps, in the southern Indo-Pacific region off the north coast of Gondwana during the Jurassic period.[23] *Sagmariasus* may have lived in the subtropical region of the southwest Pacific in the late Eocene epoch, about 35 MYA, and certainly during the Oligocene epoch 24–34 MYA,

Geological eras, periods and epochs, from the emergence of the first lobsters in the Permian period.[20] (Note, though, that the Paleozoic era extends back to about 540 MYA.) 'Tertiary' is more or less equivalent to 'Cenozoic'.

MYA	Era	Period	Epoch
0–2	Tertiary	Neogene	Pleistocene
2–5			Pliocene
5–24			Miocene
24–34		Paleogene	Oligocene
34–56			Eocene
56–65			Paleocene
65–99	Mesozoic	Cretaceous	Late
99–146			Early
146–157		Jurassic	Late
157–175			Middle
175–200			Early
200–237		Triassic	Late
237–245			Middle
245–251			Early
251–260	Paleozoic	Permian	Late
260–271			Middle
271–299			Early

as *S. flemingi* in New Zealand. When the south-flowing subtropical current subsequently retreated north as Australia migrated north, *S. verreauxi* was left in the north Tasman Sea.[24] And each of the present subpopulations of *S. verreauxi*, one on each side of the Tasman, have retained similar life histories despite their separation at least 10 million years ago. *Jasus* probably originated in the southern Atlantic-Indian Ocean region at about the same time as *Sagmariasus*, and from a common ancestor.

South African lobster biologist Dave Pollock had in 1990 offered a slightly different analysis.[25] He proposed an earlier emergence of a form ancestral to both *Sagmariasus* and *Jasus*, in the South Pacific about 60 MYA, during the Paleocene

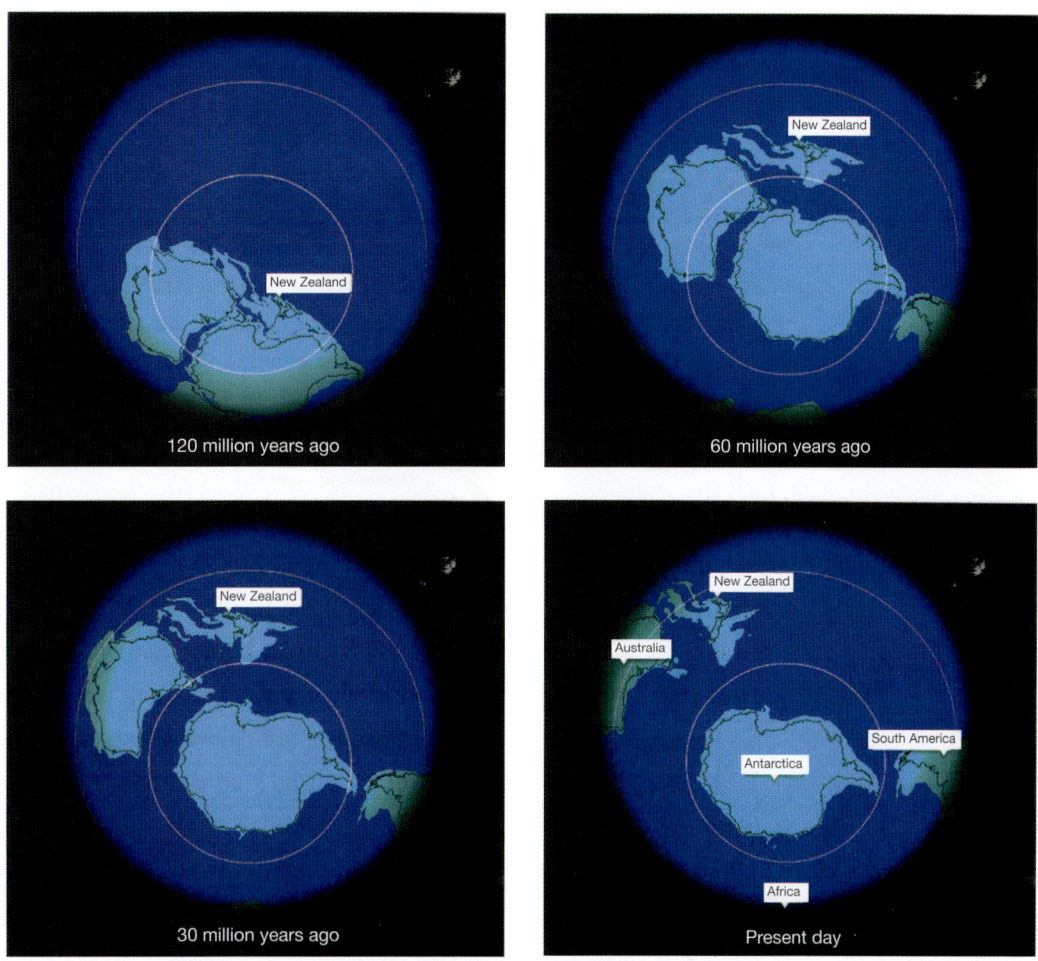

The breakup of Gondwana 120 million years ago, leading to the creation of the Tasman Sea.

epoch. The warmer-water *Sagmariasus* then evolved in the north and the cooler-water *Jasus* in the south, in the vicinity of what is now the Ross Sea. Both forms were still present in the South Pacific during the early Miocene, but with the opening of the Drake Passage at the tip of South America, and the beginning of the Antarctic Circumpolar Current, the *Jasus* form spread around the Southern Ocean.

The evolution of spiny lobsters as a group—the sequence in which the various genera and species emerged—can be hypothesised after examining and weighting body characters, genetics, behaviour, and so on. One of the most recent such phylogenies, based on 79 body characters, not unexpectedly shows packhorse forming a sister lineage with *Jasus*.[19] But in perhaps the most recent such analysis, *Sagmariasus* formed a sister lineage with the primal *Projasus*.[26] These results reinforce that the genus *Jasus* is no place for the packhorse, and that *Sagmariasus* really is primitive.

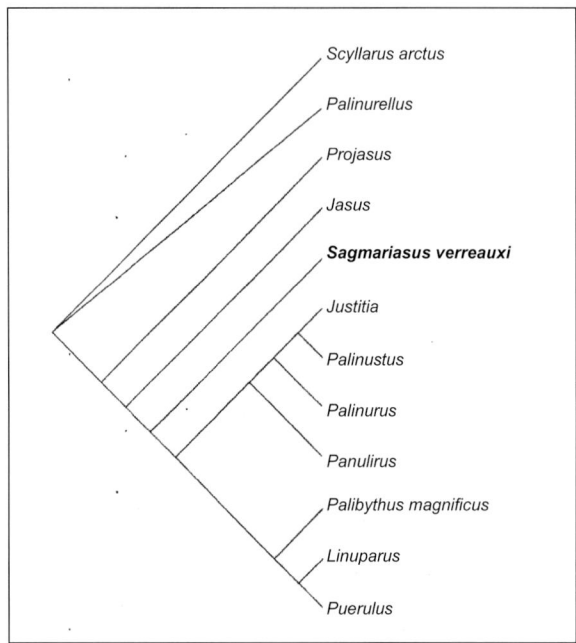

Chart of evolutionary relatedness for the spiny lobster genera based on body characters.[19] The slipper lobster *Scyllarus arctus*, being in a related but different family, was the outgroup for comparison.

CHAPTER 3

PRESENCE AND ABSENCE

Packhorse spiny lobsters distinctly prefer warmish climes. At the heart of their New Zealand distribution are the alluringly diverse shorelines of subtropical Northland. First is the 40-kilometre-wide northern tip, from Cape Maria van Diemen in the west to North Cape in the east (the 'Far North'), composed of great dune beaches interspersed with headlands of ancient volcanic rock. This exposed coast provides little respite for vessels during heavy weather that has anything but southerly in it. In contrast, the 400-kilometre east coast from North Cape south into the Hauraki Gulf is protected from the prevailing westerly wind and contains some of the most magnificent shoreline scenery on Earth. There are harbours—Whangaroa, the Bay of Islands, and the Hauraki Gulf itself—capable of concealing a whole flotilla of ships. Many smaller but still substantial havens give a sort of jigsaw pattern to this coast: Parengarenga, Houhora, Whangamumu, Whangaruru, Whangarei, and Mahurangi. The shoreline here is characterised by massive cliffs of old sedimentary rock alternating with beaches of rolling dunes and intimate coves. There is also a generous scattering of islands, both inshore and offshore, large and small.

These north and east coasts of Northland are the shores most often associated with packhorse, but in fact the species is far more widespread.[1] Sizeable stocks of breeding lobsters have, sadly, been fished out all the way from Great Barrier Island, off the tip of Coromandel Peninsula, to White Island in the eastern Bay of Plenty. On the east side of Coromandel Peninsula the harbours are typically smaller than those further north, but the terrain is no less rugged or varied.

Further south the topography mellows somewhat into the long sandy stretches of the Bay of Plenty—but with sentinel volcanic peaks onshore, offshore and underwater leaving no doubt that this is mainland New Zealand's northern contribution to the Pacific's tectonic ring of fire. South of East Cape the numbers of breeding packhorse plunge, but the youngsters are still to be found all the way to Cook Strait. The sedimentary rocks here are young too, the coastal waters accordingly taking on a more murky appearance. What particularly characterises this region are rivers which, when in flood, carry huge loads of sediment. The Waipaoa, Waiapu and Hikuwai rivers, whose catchments make up merely 0.003 per cent of the Earth's land-surface area, collectively deliver 0.3 per cent of all of the mud that flows into the Earth's oceans. Poor land use upstream, in the face of a vigorous maritime climate and land instability, is the main cause of this astonishing output.[2] The remote Wairarapa coast reaches south to Cook Strait and is topographically too tortuous to own a continuous coastal road. Many of its tiny picturesque settlements are no more than 100 kilometres from the centre of Wellington yet are many, many hours by road.

Although this journey has covered well over 1000 kilometres of shoreline, packhorse are actually a rather localised species in a global sense. Known only from eastern Australasia—the waters of the west South Pacific Ocean and the Tasman Sea—they join a dozen or so other spiny lobsters around the world that occupy subtropical waters.[3] Today's distribution of this and other lobsters is the product of several forces. First and foremost has been the Earth's geological history and the way that species evolved within its ever-changing seas (see Chapter 2). Presently, the distribution of packhorse reflects larval drift, together with the migrations of the juveniles and, from time to time, the wanderings of an occasional adult.

Breeding populations of packhorse are found in waters with typical surface temperatures of about 16°C in winter and 22°C in summer, but the juveniles are much more cosmopolitan. In New Zealand, packhorse are most abundant along the north and east coasts of the North Island, but captures are not uncommon along the entire west coast of the country. The species has also been known as an occasional straggler on the northeast coast of the South Island, at the Kermadec Islands to the northeast, at the Chatham Islands to the east, and off the west coast of Stewart Island in the south. In Australia, *Sagmariasus verreauxi* is most common in New South Wales. But its range is broader than this, being found

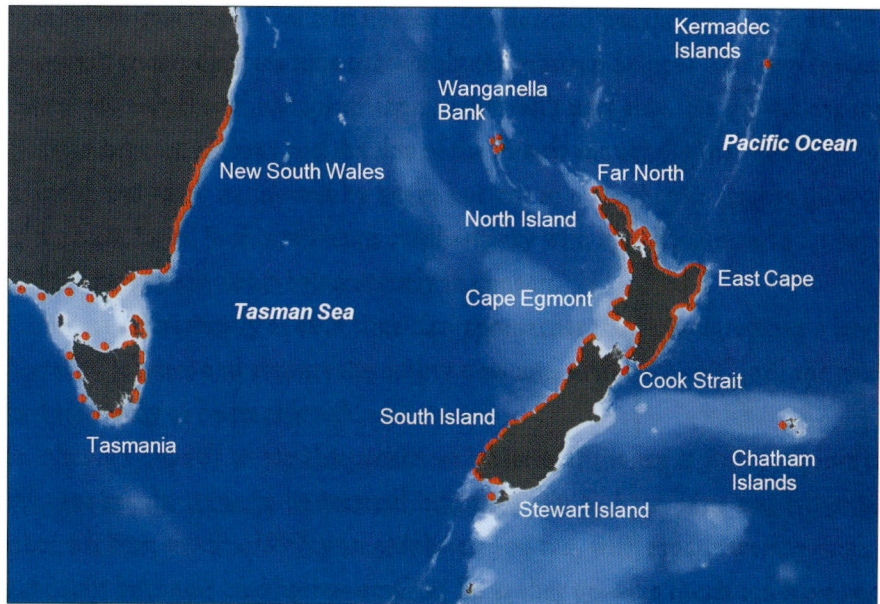

The world distribution of the packhorse spiny lobster.[1] Solid lines show main distribution; dots show isolated individuals.

along the east coast from Tweed Heads, just south of the New South Wales–Queensland border, south to Tasmania, and then as far west as Port MacDonnell in South Australia.[4] The most northern location for packhorse is probably the Wanganella Bank, 500 kilometres northwest of Cape Reinga, where a commercial vessel caught several sacks of mainly small individuals during a short spell of fishing in the late 1960s.[1]

Although packhorse are confined in their distribution, the same cannot be said of their *Jasus* cousins. Six members of the genus are scattered widely in southern waters—including on seamounts and ridges. Those at Tristan da Cunha live at what is said to be the most remote inhabited location on Earth.

Ocean envelops more than two thirds of the surface of Earth to an average depth of almost 4 kilometres,[5] and—just like the land surface—its floor is anything but uniform. Intermittent soundings across ocean basins during the 19th and early 20th centuries—taken by hand and depicted on charts as lonely lines, often emanating from ports, like the radiating strands of a spider web—were expanded during the mid 20th century as electronic sounders on survey vessels provided access to more detailed deep-water bathymetries. Diversified shipping

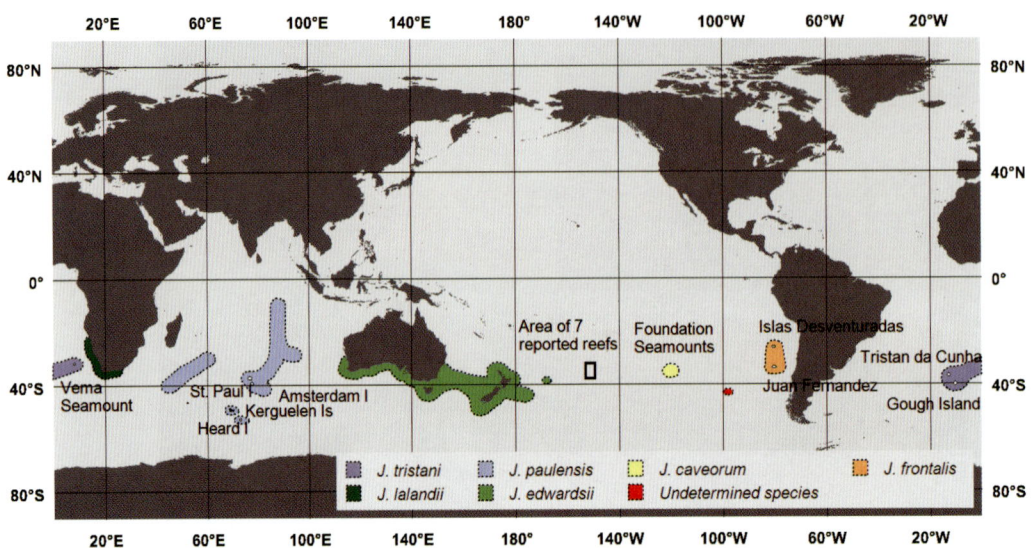

Indicative distribution of the six known (and one undetermined) *Jasus* species, together with the location of places referred to in this chapter.[9,15] There is every chance that *Jasus* is more widespread in the east South Pacific Ocean than indicated, and *Jasus paulensis* is probably to be found from time to time north of the Equator.[9]

routes helped fill out this detail. Satellite-derived gravitational anomaly data, together with information on how the surface level of the ocean varied spatially, then led to synoptic peeks of seafloor shape on the scale of ocean basins, and accurate depictions of those seamounts and ridges close to the surface. Presently swath mapping using multibeam echosounders reveals features just a few metres in size. Now for the first time the shape of the seafloor is being seen in the same detail as the land.

Seamounts and rises are of great interest, not only to fishing companies who focus on the rich resources often associated with them, but also to taxonomists and ecologists because of the creatures that live in this poorly known, often dangerously sulphurous environment. Many are even potential mines for valuable minerals. Seamounts can be defined as seafloor features with a vertical profile of at least 250 metres, but they vary enormously in size and shape, and their height above the seafloor and depth below the surface of the water. From the 1960s to the end of the 1980s, the former Soviet Union led the exploration of seamounts—hundreds were examined as potential fishing grounds in all the great oceans.[6] Now many other nations are into seamount exploration. And it is clear that the shallower of the seamounts, ridges and rises, particularly those in the higher southern latitudes that reach within 500 metres or so of the surface, can be important homes to spiny lobsters.

One of the first spiny lobster species to be fished commercially on a non-emergent seamount was the Tristan rock lobster *Jasus tristani*. Known previously only from the remote Tristan da Cunha group of islands, this lobster was found in abundance in 1964 on the almost as far-flung Vema Seamount, 2000 kilometres to the northeast.[7] The hectic couple of years of international free-for-all that followed fished them out.

Later, an area of the South Pacific Ocean almost the size of New Zealand, 3000 kilometres east of the North Island and 2000 kilometres south of Tahiti, began to tantalise lobster fishers down under. Seven shallow reefs had been reported there, the first being the Maria Theresa Reef, observed in 1843 from a vessel of the same name.[8] Later, in 1899, the *Wachusett* passed over an area of dark grey with deep blue on either side, apparently a coral reef 9–11 metres beneath the surface—Wachusett Reef. The reef most recently reported was marked by breakers seen in 1963.

The Maria Theresa Reef received special attention. As Jim Eade of the then New Zealand Oceanographic Institute recounted:

> In 1966 three photographs were published in the radio amateur's journal 'CQ' of DX radio enthusiast Don Miller operating from a reef given to be Maria Theresa. The only other information given on the locality is the DX call prefix F08M which is DX zone 0-32, French Oceania. The photographs show small rocky patches, distinctly coral-like in appearance, standing no more than 1 m above sea level. These seemed to be scattered over an area at least 100 m across.[8]

Accordingly, searches were arranged for this and the other six reefs, which might menace shipping but also—more importantly for several involved—might harbour spiny lobsters. One of the first off the mark was Wellington trawlerman and lobster fisherman Alan Aberdein, in 1972. The noon position of MV *Picton* on 29 February was 12 kilometres from the reported position, the day fine and clear. A 4-hour search with an echo sounder was made of an area 16 by 16 kilometres centred on the reef's position, but no bottom was recorded and no sign of any reef seen.

Seven shallows were reported south of Tahiti, the intrigue heightened by this image given to be Don Miller operating a radio from Maria Theresa Reef (Don and the radio equipment are sitting on the reef itself—more of the reef can be seen as dark patches in the background).[8]

Not one of these seven reefs was ever found—because they don't exist. The waters here are kilometres deep. Jim Eade explained why so many reefs should have been reported from just one relatively small area. Although the number and close proximity would seem to support the existence of at least one reef, the position(s) of which had been only crudely fixed, it is also likely that one mistaken sighting increased the chance of more mistaken sightings in the same area. After the first report, mariners would have kept a sharp lookout for submerged dangers in the vicinity. And it seems true that the area is one with strange oceanographic goings on. Long after Alan Aberdein's foray, lobster explorer John Chadderton came across, in the same area, a depression in the sea surface seven or eight metres deep and several hundred square metres in area.[9]

But not all offshore searches for spiny lobsters have been so utterly without success. In 1988, John Chadderton revealed the presence of red rock lobsters *Jasus edwardsii* on shallow banks in the west Tasman Sea.[10] Another relative of the packhorse and a fellow Silentes, the Chilean jagged lobster *Projasus bahamondei*, is potted on seamounts and rises on the Nazca Ridge off northern Chile.[11] Its biomass has been estimated to be in the thousands of tonnes. The closely related Cape jagged lobster *Projasus parkeri* is widely found on seamounts and rises in the west South Pacific and in the south Indian Ocean,[12] sometimes in great numbers.[9] Also found in quantity in the Indian Ocean, south of Madagascar, are the deep-water lobsters *Palinurus delagoae*[13] and *Palinurus barbarae*,[14] and, in shallower waters, *Jasus paulensis*.[15]

And this leads to one of the most intriguing episodes in the history of spiny lobster exploration. It concerns the St. Paul rock lobster *Jasus paulensis*.

'Chirac says: "I ordered Navy to fire"' is a headline in Wellington's *Evening Post* on 20 October 1986. Jacques Chirac, Prime Minister of France, had personally given the order to fire on the *Southern Raider*, a converted Japanese tuna-poling vessel with 23 aboard (four New Zealanders, one Swede, and half a dozen Australians; the rest Korean lobster-processors) and, if necessary, to sink it. According to Chirac, the *Southern Raider* was allegedly fishing illegally for lobsters inside French waters and had failed to heed warnings or allow a search from a naval patrol after being intercepted off the French island of St. Paul in the central south Indian Ocean. 'It sank some two hours after the *Albatros* opened fire with a machinegun and then with a 40 mm gun.'

A few days previous the newspaper had been more specific, saying that 'five shells were fired at the trawler after it had tried to flee and had refused to answer radio calls' and that the sinking followed more than 30 hours of warnings.[16] Among the crew, now not heard of for several days, was New Zealander Arthur Symes. His wife had trouble answering their children when they asked 'Has daddy been blown up?'[17] Fortunately, most of the crew was soon free on Réunion Island, almost 3000 kilometres to the northwest, and then on their way home. Not so fortunate were either the captain or the chief officer.

The Australian captain of the *Southern Raider*, John Chadderton, has long been captivated by the existence of spiny lobsters in remote places, and especially on seamounts. This interest has taken him to such far-flung places as the waters around Pitcairn Island, the Louisville Ridge east of New Zealand, the Gascoyne Seamount in the Tasman Sea, and especially the south Indian Ocean. The catch of St. Paul rock lobsters from the 1986 *Southern Raider* voyage, most from huge round pots, was destined for Japan. But while sheltering in the lee of St. Paul Island to undertake structural repairs required after a particularly vicious storm to the south, the crew became aware of a large, light-coloured shadowing vessel. It did not carry any flag of nationality, nor did it respond to frequent calls on the international radio frequency 2182 or to light signals. After 9 hours, and soon after the unidentified vessel had hoisted flags to signal that they wished to communicate, 40-millimetre cannon shells were fired into the water around the *Southern Raider* and machine gun fire destroyed their radar and radio antennas. 'It was considered that this action and procedure was hostile and very suspicious,' recounted John Chadderton with remarkable detachment.[9]

Events came to a head when, at 6 pm on 9 October 1986 and at a point 250 kilometres southwest of St. Paul Island, signal flags were again hoisted saying 'Stop or I will fire on you'. But John wasn't going to stop now for a vessel that wouldn't identify itself or declare its reason for wanting him to halt here in international waters, especially after having been attacked by it earlier in the day. The vessel bombarded them with explosive incendiary armour-piercing shells and heavy machine-gun fire, with devastating effect. It was a miracle anyone survived the onslaught. With holes quarter of a metre wide below the water line and the engine disabled by a shell, the *Southern Raider* drifted to a stop, down by the stern and listing heavily to starboard. Life jackets were issued and the surviving life rafts launched.

The attacking ship was the French naval patrol vessel *Albatros*, a converted trawler close to 100 metres in length. Its Zodiac boat eventually took the *Southern Raider*'s life rafts under tow and the crew was delivered to the French island of Réunion. However, they did not land there until tense negotiations with a French secret service agent flown out to the *Albatros* by military helicopter revealed their bleak prospects unless they admitted to activities of which they were no part—espionage. This they were not prepared to do.

Dramatic image of the *Albatros*' engagement with the *Southern Raider* before she was shelled and sunk on 9 October 1986 in the south Indian Ocean.

At War with the French, a Sixty Minutes documentary broadcast on Australia's Channel 9 on 15 February 1987, recounts some of the events that followed. It begins with John Chadderton being led handcuffed from the *Albatros* in Réunion. Later he describes to camera how his vessel had been illegally sunk, the broadcast even showing some of that action, apparently captured surreptitiously by a crewman on the *Albatros*. Newspapers of the time reported that John and Chief Officer Alistair Annandale had each received 6-month prison sentences, together with $10 000 fines, for illegally fishing French waters and avoiding arrest.[18] Both John's account, and the Sixty Minutes documentary, had the pair of them being treated by the prosecutor more or less as enemies of France simply because they were Australian citizens.

The two appealed their sentences and during the following weeks were more or less free to roam the island—unencumbered by their passports. It was suggested in the documentary, and in newspapers of the time,[18] the real reason that the vessel had been sunk by the *Albatros* was that it had strayed too close to the Kerguelen Islands. These islands had apparently been earmarked by the French as a possible replacement site for their hugely unpopular nuclear testing activities at Mururoa Atoll in the South Pacific. Notably, France had reacted to protests against these activities by bombing the Greenpeace protest ship *Rainbow Warrior* in Auckland Harbour in 1985. Did the French sink the *Southern Raider* for suspected intelligence gathering, using the pretence that she had been fishing illegally?

John Chadderton and Alistair Annandale then effected a remarkable escape. Next interviewed in Singapore, John circumspectly disclosed that 'Well, we sort of swam and windsurfed the major portion of the distance' from Réunion to Mauritius. It took them 18 hours to make the 200-kilometre, open-ocean passage.

While researching these events, things got curiouser and curiouser. The *Southern Raider* had among its crew the Swede Johan Salén, who wrote an account in Swedish of their experiences being sunk by the *Albatros* and confined on Réunion Island. His book *Sänkta*, published in 1987 by Bonniers, Stockholm—a not insubstantial publishing house—could not in 2007–08 be recovered by title or author using regular internet search engines. It was finally tracked down through the more powerful tools of the Museum of New Zealand Te Papa Tongarewa library—to the *fiction* shelves of the British Library. When we requested the book as an interloan, as Te Papa routinely does—receiving even extremely rare and valuable tomes—the British Library revealed after some delay that it was not available for loan overseas. It was almost as if someone did not want this book read too widely at all. After having eventually obtained a second-hand copy of *Sänkta* in Sweden, and having had it read by native Swedish-speaker Kristina Hjelm of

Russell, it became clear that this book was anything but fictitious. Instead, it provided a blow-by-blow account of what were almost certainly actual events.

The sinking of the *Southern Raider* and the subsequent events on Réunion Island made headlines in France, New Zealand, Australia and Sweden. And there is no evidence that John Chadderton was illegally fishing French waters, as later acknowledged by a somewhat remorseful French Defence Minister André Giraud.[18] Two weeks of searching had failed to find any fishing gear in the vicinity of St. Paul Island—because John's pots were in waters way to the south when attacked.[9] Undeterred, and with compensation in hand, John Chadderton went on to fish vast tracts of the Indian Ocean. His passion to discover and fish remote bastions of spiny lobster distribution over many years has without doubt greatly enlarged the known distributions of spiny lobsters in the region, and enlightened us about their ecology. He showed that the St. Paul rock lobster, thought from its description in 1862 to be confined to a couple of tiny mid-ocean islands, is in fact to be found over a region of the south Indian Ocean not that far short of the size of Australia. It extends further south than any other spiny lobster—to the Kerguelen Islands (49°S) and Heard Island (53°S). And, remarkably, it appears to reach its northern limit north of the Equator: '. . . This will surprise you as they [St. Paul rock lobsters] were taken in 260 mtrs depth off the Nicobar Islands at 7 deg lat North . . .'[9] A temperate species straddling the Equator? At first reading this seems fanciful. Surely the species has been confused, or there has been a memory lapse regarding location. But there is no reason to suspect John's account of its presence north of the Equator. He certainly knew a St. Paul rock lobster when he saw one, having handled multitudes of them over the years; and, working in such remote places with no easy points of reference, he had to be meticulous in his position- and record-keeping. Interestingly, decades earlier, early-stage *Jasus* larvae were reported to have been caught off the coast of India, inspiring a conversation about a Northern Hemisphere presence for the genus that only now really seems to have legs. The St. Paul rock lobster is therefore probably the only species of *Jasus* to be found either on the Equator or north of it, but its presence there is likely to be ephemeral, depending on the vagaries of the currents that transport the larvae.

An intriguing visitor to John's pots in the cool southern waters was a *Jasus* lobster smaller and darker than his regular St. Paul ones. It had short, thin legs;

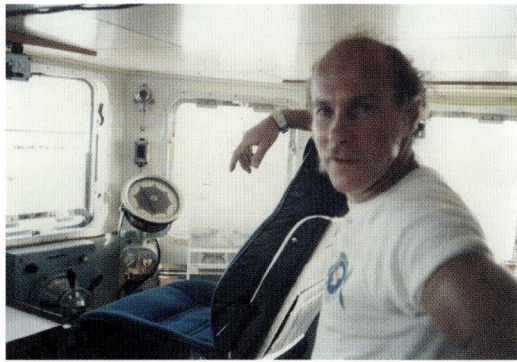 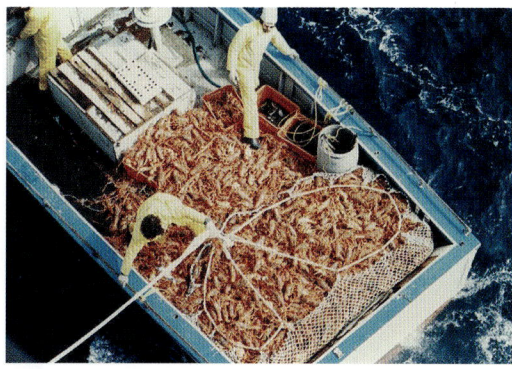

The St. Paul rock lobster was known only from around tiny St. Paul Island and the nearby Amsterdam Island until John Chadderton (left) potted them, often in great numbers (right), over a broad area of the central Indian Ocean.

a wider and shorter tail with more outwardly projecting spines; and shorter feelers. Reaching only 300 grams, this lobster formed about one tenth of catches around Kerguelen and Heard Islands, and it was also found elsewhere south of St. Paul Island.[9] A curious feature of this lobster was that it survived being frozen to −40°C, providing the freezing was carried out in stages. Thawed, it resumed normal movement. Like certain insects and cold-water crabs, these lobsters presumably contain antifreeze proteins in their tissues, but their status as a species is yet to be formally determined.

Some of John Chadderton's catches of the St. Paul rock lobster from within its main area of distribution were phenomenal. Using pots 2 metres in diameter and a metre high, attached to their floats with 40-millimetre rope, one-tonne lifts of this generally small lobster were not uncommon: '. . . the best we ever did was 600 tons of live and 172 tons of whole cooked and green [whole uncooked] frozen for 156 days . . . Mind you this is working about 18 to 21 hr days to achieve this and I can tell you the crew wished they would never see a lobster again and would pray for empty pots.'[9] Clearly not only are these spiny lobsters widespread, but in places incredibly profuse.

John also showed us that the Cape jagged lobster *Projasus parkeri*, until recently known from the Indian Ocean only through small numbers taken in deep waters directly off the east coast of South Africa, is in fact common and widespread in the Indian Ocean.[9] Fishing to depths of 3500 metres, John found them from 300 to 3200 metres—possibly the widest depth range for any crustacean. In places

John Chadderton caught what appeared to be two distinct forms of *Jasus* in the Indian Ocean (left). The Cape jagged lobster is the deepest-living of all spiny lobsters and has most often been found associated with seamounts—in this case one recently volcanically active, given the look of the seafloor.

they were so abundant that pots arrived at the surface containing 600 individuals. Bright orange in the shallower waters, at intermediate depths these lobsters were pink, and the deepest ones were pearly white.

Another of the packhorse's cousins is of immense interest. It was the first new species of *Jasus* discovered for more than a century. And it wasn't just a single specimen—there were thousands.

Joe Cave, a lobster fisherman from New Zealand's Stewart Island, had obtained satellite data which indicated a chain of seamounts close to the surface in the middle of the South Pacific Ocean. He dispatched his 28-metre vessel *David Baker* in mid 1995, skippered by Steven King. After three weeks they reached the Foundation Seamount Chain, 2000 kilometres southeast of Pitcairn Island and 4000 kilometres from Chile. They potted 12 000 lobsters.

This spiny lobster was clearly a *Jasus* but, unlike all others in the genus, it had very little sculpturing on the upper surface of its tail. Rick Webber and I named it *Jasus caveorum*—acknowledging Joe Cave's interest in finding lobster populations in remote locations, and his brother Ernie's contribution to developing the red rock lobster fishery in remote parts of southern New Zealand and his pioneering work there into escape gaps that allow small lobsters to readily leave pots.[19] The specific name *caveorum* is also a play on the word 'cave', implying a hidden species with no surface indication of its presence, for this lobster is known (so far) from only this one submerged seamount chain. Later, several others mounted

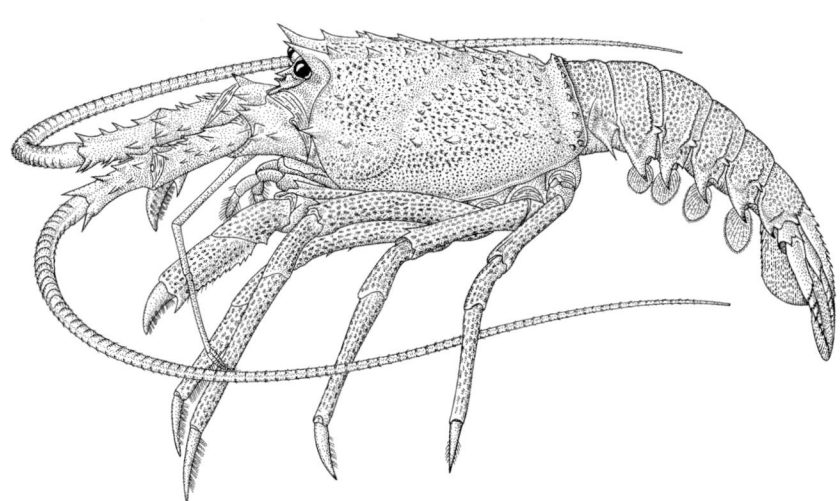

Joe Cave's party fished the Foundation Seamounts rock lobster, the first new species of *Jasus* discovered in more than a century, on a remote seamount chain in the east-central South Pacific. The surface of its tail is almost smooth, as illustrated by Rick Webber.

expeditions to pot the spiny lobsters on this remote spot in international waters, and there was also damaging bottom trawling for orange roughy.

But by no means is that all there is concerning spiny lobsters in remote parts of the South Pacific Ocean. During the late 1980s, John Chadderton extensively surveyed great tracts of ocean between New Zealand and Chile, and particularly in the region of Pitcairn Island. He potted spiny lobsters on several seamounts, which likely included those of the Foundation Seamount Chain.[9] John's records are boxed up in storage so full details are yet to be revealed, but it would be extraordinary—given what we know of the distribution of other *Jasus* species, and their long larval lives—if *J. caveorum* was confined to one small set of seamounts. Tantalisingly, an as-yet-unidentified *Jasus* lobster was taken by the tonne in the early 1960s on a seamount 2000 kilometres southeast of the Foundation Seamount Chain.[15] There is every chance, therefore, that South Pacific stocks of spiny lobsters—some possibly huge—await enterprises with pockets, vessels and fuel tanks capacious enough to work grounds thousands of kilometres distant from any port.

And this still leaves the South Atlantic Ocean: a *Jasus* larva, identified through its form and DNA but unable to be associated with any known species, was taken in the South Atlantic Ocean in the early 1990s.[15] Nature, particularly at its fringes, clearly has more to reveal . . .

Earlier in this chapter was a chart showing the world distribution of *Jasus* spiny lobsters as we presently know it. Preparing maps of this type is fraught with

New Zealand's spiny lobsters. Upper left, the red rock lobster *Jasus edwardsii*, the most widespread and common species. Upper right, the packhorse *Sagmariasus verreauxi*, the other commercial species. Lower left, the deepwater (or Cape jagged) lobster *Projasus parkeri*, seamount footage of them being suggestive of commercial-sized stocks. Lower right, the white-whiskered spiny lobster *Panulirus femoristriga*, occasionally reported at the Kermadec Islands.

difficulties, mainly the scale of the map, discontinuities in lobster occurrence, and the lack of detailed distributional information on lobsters living in offshore areas. The appearance from time to time of still-to-be-confirmed occurrences, where the species of lobster have not been categorically determined, means that maps such as these are always works in progress. But no matter how incomplete and preliminary, they are still of immense interest to zoogeographers and ecologists, and, of course, to lobster fishers.

Indicative maps of distribution for all the other spiny lobsters follow. There are two fairly well-marked patterns of distribution among the genera.[23] First is the low-latitude global distribution of *Panulirus*—confined to the shallows; and, in deeper waters, the somewhat more fragmentary appearance around the globe of *Justitia, Linuparus, Nupalirus, Palibythus, Palinurellus, Palinustus,* and *Puerulus*. Then there are the higher-latitude genera *Jasus, Projasus, Palinurus,* and *Sagmariasus*, the first two having circumpolar distributions. Spiny lobsters have not been reported north of 60°N (the European spiny lobster *Palinurus elephas* in the north of Scotland is the northernmost) or south of 53°S (the St. Paul rock lobster, reported by John Chadderton to be at Heard Island in the south Indian Ocean). In addition to *Sagmariasus*, three other genera—*Jasus, Projasus* and *Palibythus*—are, to all intents and purposes, confined to the Southern Hemisphere, all others being found both north and south of the Equator. Species confined to the south are a third again more numerous than those found only in the north, a pattern common among marine crustacea.[24]

Seldom in any one place does more than one genus occur, and where they do they almost always live in different habitats.[25] The latter also applies at the species level. The Indo-West Pacific has more species of spiny lobster, and indeed of most crustaceans, than anywhere else. India, for example, has six species of *Panulirus* and one each of *Puerulus, Linuparus* and *Palinustus*. This richness is brought about by the geographic location, the presence of both subtropical and tropical conditions, and the range of habitats, from mud and sand through to rock and coral reef.[26] In the Atlantic Ocean, the highest species diversity is found off the coast of Central America, whereas in the southwest there are no spiny lobsters at all. In the Pacific Ocean, there are many spiny lobsters in western and central areas, but just two are found in the southeast, and of these two, only *Projasus bahamondei* reaches the coast of South America.

Depth ranges vary enormously. Most *Panulirus* species are found in very shallow waters, the scalloped spiny lobster *Panulirus homarus*, among others, even being captured on coral-reef tops at low tide. Contrast this with the Cape jagged lobster *Projasus parkeri*, which has been taken from waters more than 3 kilometres deep. The great depths in which this species lives allow for wide latitudinal distribution because temperatures at depth are much more uniform than they are in the shallows. For example, at a depth of 500 metres, the waters of the South Pacific Ocean are a fairly constant 7°C all the way from the Equator to 50°S.[27]

Although the discovery of new genera of spiny lobsters is unlikely, there is every chance that further spiny lobster stocks—including new species—will be discovered as the vast unexplored areas of remote or deep habitats in oceanic waters are sampled for the first time. We have seen this with *Jasus* in the Indian and South Pacific oceans. Although they may be densely distributed within a particular location, populations of shallower water species in remote areas are likely to be smaller than populations of species associated with large land masses where there is plenty of habitat. They probably built up slowly on seamounts and ridges over millennia, their levels of recruitment variable and depending on the vagaries of the currents of the high seas. They are therefore unlikely to be able to support sustained heavy fishing, if experience to date is anything to go by. On the other hand, stocks of such deepwater genera as *Projasus* and *Puerulus* may be both abundant and widespread. But, as fishing them is technically challenging, they are very much an unknown quantity: their populations are difficult to assess for size and productivity, and the manner of their recruitment is poorly understood.

This chapter closes with illustrations of New Zealand's slipper lobsters (scyllarids), or at least the ones we know about; almost certainly there are more species to be revealed. Aotearoa's dominant species are the diminutive New Zealand slipper lobster *Antipodarctus aoteanus*, the larger prawn killer *Ibacus alticrenatus*, and the commercial-sized Spanish lobster *Arctides antipodarum*. None of them are commercially fished here.

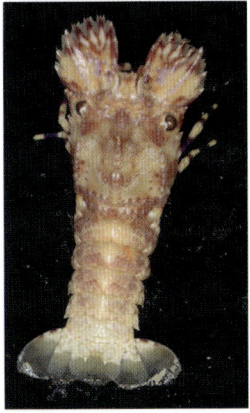
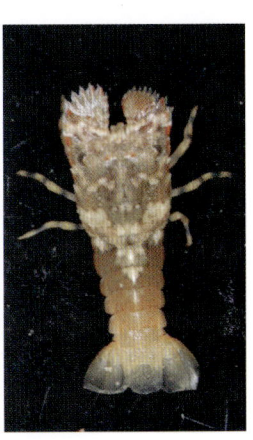

New Zealand's slipper lobsters.
Upper from left: its smallest, the New Zealand slipper lobster *Antipodarctus aoteanus* (to 7 cm total length) and its largest *Scyllarides haanii* (50 cm), both being found in our warmer waters. Middle from left: the widespread prawn killer *Ibacus alticrenatus* (20 cm), *Ibacus brucei* (20 cm) and *Antarctus mawsoni* (8 cm) from the Kermadec Islands, and an undescribed scyllarine (8 cm) from north of New Zealand. Lower: the Spanish lobster *Arctides antipodarum* (30 cm), living in northern waters.

Earth's spiny lobster species. Silentes genera are highlighted in bold; the rest belong to the Stridentes. Common names, where given, are based (not exclusively) on Lipke Holthuis' catalogue.[3] Note that *Panulirus argus* appears to be composed of two subspecies, *Panulirus homarus* of three, and *Panulirus longipes* of two.[28]

Genus	Species	Common name
Jasus	caveorum	Foundation Seamounts rock lobster
	edwardsii	Red spiny/rock lobster
	frontalis	Juan Fernández rock lobster
	lalandii	Cape rock lobster
	paulensis	St. Paul rock lobster
	tristani	Tristan rock lobster
Justitia	longimanus	
Linuparus	somniosus	African spear lobster
	sordidus	Oriental spear lobster
	trigonus	Japanese spear lobster
Nupalirus	chani	
	japonicus	
	vericeli	
Palibythus	magnificus	Musical furry lobster
Palinurellus	gundlachi	Furry lobster
	wieneckii	Indo-Pacific furry lobster
Palinurus	barbarae	Madagascar Ridge spiny lobster
	charlestoni	Cape Verde spiny lobster
	delagoae	Natal spiny lobster
	elephas	European/common spiny lobster
	gilchristi	Southern spiny lobster
	mauritanicus	Pink spiny lobster
Palinustus	holthuisi	Asian blunthorn lobster
	mossambicus	Buffalo blunthorn lobster
	truncatus	American blunthorn lobster
	unicornutus	Unicorn blunthorn lobster
	waguensis	Japanese blunthorn lobster

Genus	Species	Common name
Panulirus	*argus*	Caribbean spiny lobster
	brunneiflagellum	Aka-ebi
	cygnus	Western rock/spiny lobster
	echinatus	Brown spiny lobster
	femoristriga	White-whiskered spiny lobster
	gracilis	Green spiny lobster
	guttatus	Spotted spiny lobster
	homarus	Scalloped spiny lobster
	inflatus	Blue spiny lobster
	interruptus	California spiny lobster
	japonicus	Japanese spiny lobster
	laevicauda	Smoothtail spiny lobster
	longipes	Longlegged spiny lobster
	marginatus	Hawaiian spiny lobster
	ornatus	Ornate spiny lobster
	pascuensis	Easter Island spiny lobster
	penicillatus	Pronghorn spiny lobster
	polyphagus	Mud spiny lobster
	regius	Royal spiny lobster
	stimpsoni	Chinese spiny lobster
	versicolor	Painted spiny lobster
Projasus	*bahamondei*	Chilean jagged lobster
	parkeri	Cape jagged/deepwater lobster
Puerulus	*angulatus*	Banded whip lobster
	carinatus	Red whip lobster
	sewelli	Arabian whip lobster
	velutinus	Velvet whip lobster
Sagmariasus	*verreauxi*	Packhorse/eastern spiny/rock lobster

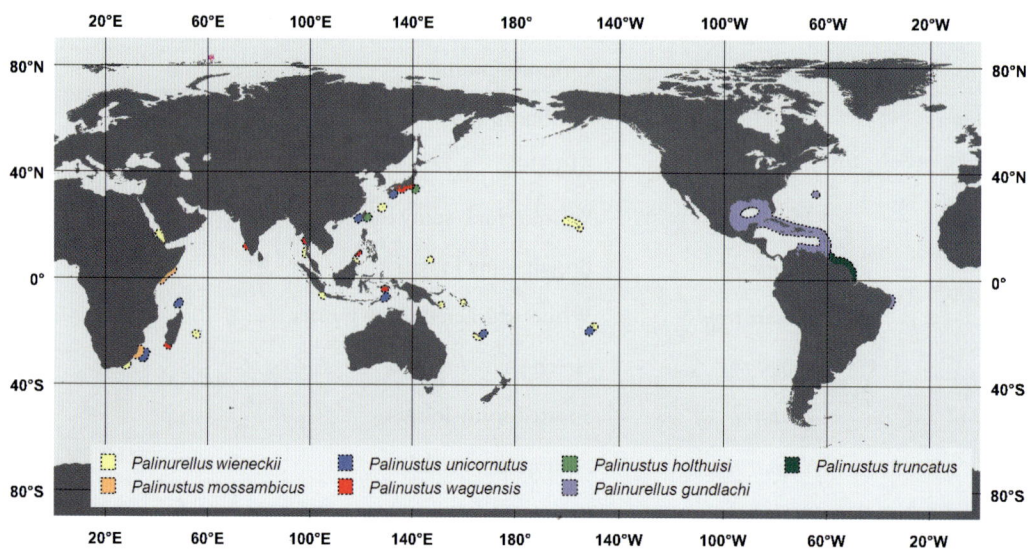

Indicative distribution of the Indo-Pacific furry lobster *Palinurellus wieneckii* and the furry lobster *Palinurellus gundlachi*;[3,29,30] and the buffalo blunthorn lobster *Palinustus mossambicus*, the unicorn blunthorn lobster *Palinustus unicornutus*, the Japanese blunthorn lobster *Palinustus waguensis*, the American blunthorn lobster *Palinustus truncatus*, and the Asian blunthorn lobster *Palinustus holthuisi*.[3,29,31]

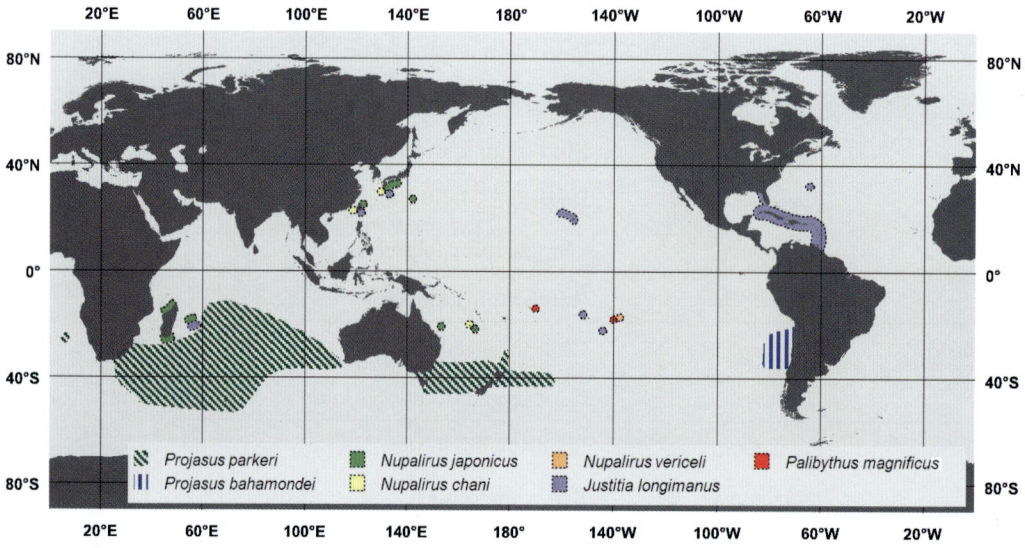

Indicative distribution of the Cape jagged lobster *Projasus parkeri* and the Chilean jagged lobster *Projasus bahamondei*;[3,9,11,12] *Nupalirus japonicus*, *Nupalirus chani*, *Nupalirus vericeli* and *Justitia longimanus*; [3,30,32] and the musical furry lobster *Palibythus magnificus*.[3]

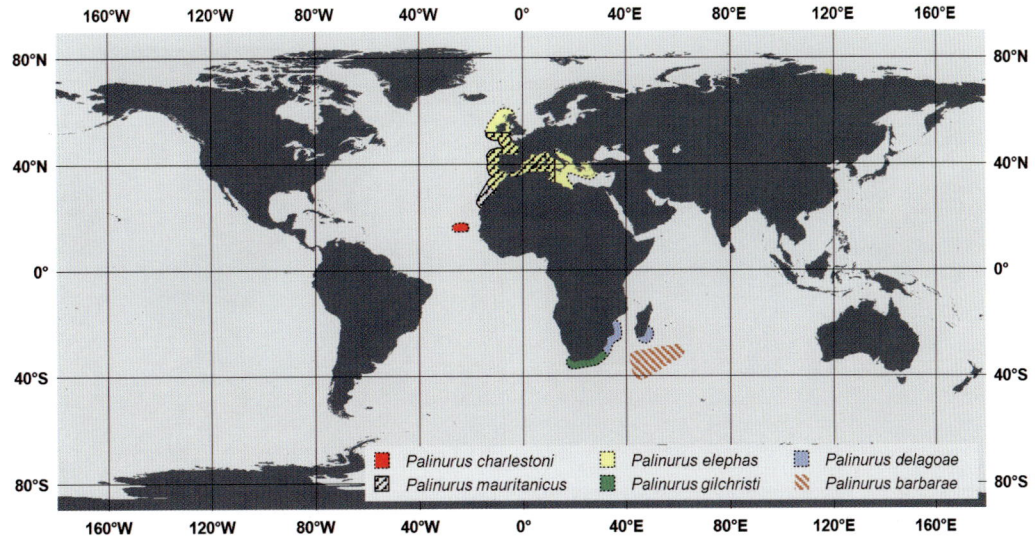

Indicative distribution of the Cape Verde spiny lobster *Palinurus charlestoni*, the pink spiny lobster *Palinurus mauritanicus*, the European spiny lobster *Palinurus elephas*, the southern spiny lobster *Palinurus gilchristi*, the Natal spiny lobster *Palinurus delagoae*, and the Madagascar Ridge spiny lobster *Palinurus barbarae*.[3,13,14]

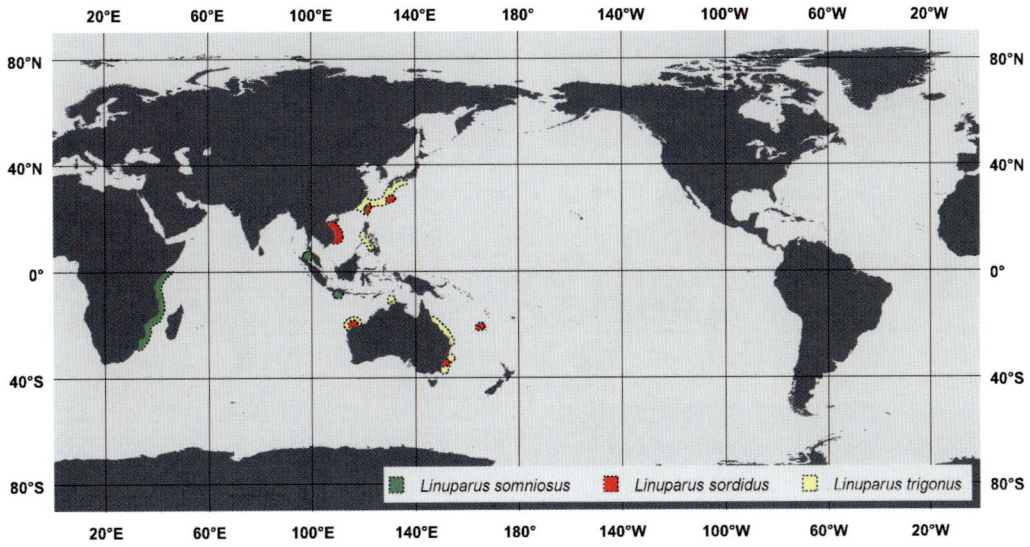

Indicative distribution of the African spear lobster *Linuparus somniosus*, the Oriental spear lobster *Linuparus sordidus*, and the Japanese spear lobster *Linuparus trigonus*.[3,29,33–37]

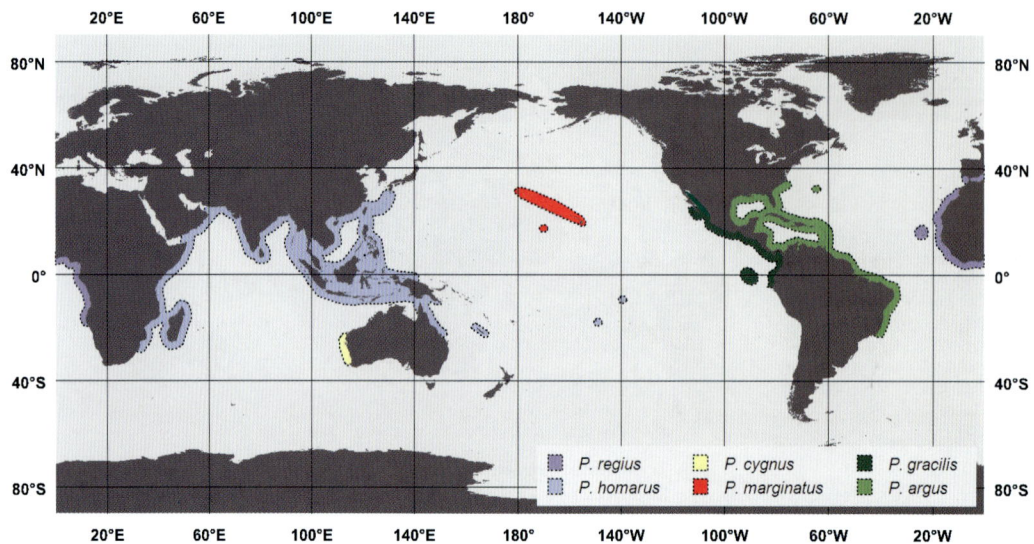

Indicative distribution of the royal spiny lobster *Panulirus regius*, the scalloped spiny lobster *Panulirus homarus* (as three subspecies), the western rock lobster *Panulirus cygnus*, the Hawaiian spiny lobster *Panulirus marginatus*, the green spiny lobster *Panulirus gracilis,* and the Caribbean spiny lobster *Panulirus argus* (as two subspecies).[3,29,38] The Caribbean spiny lobster has also occasionally been reported off the coast of west Africa.

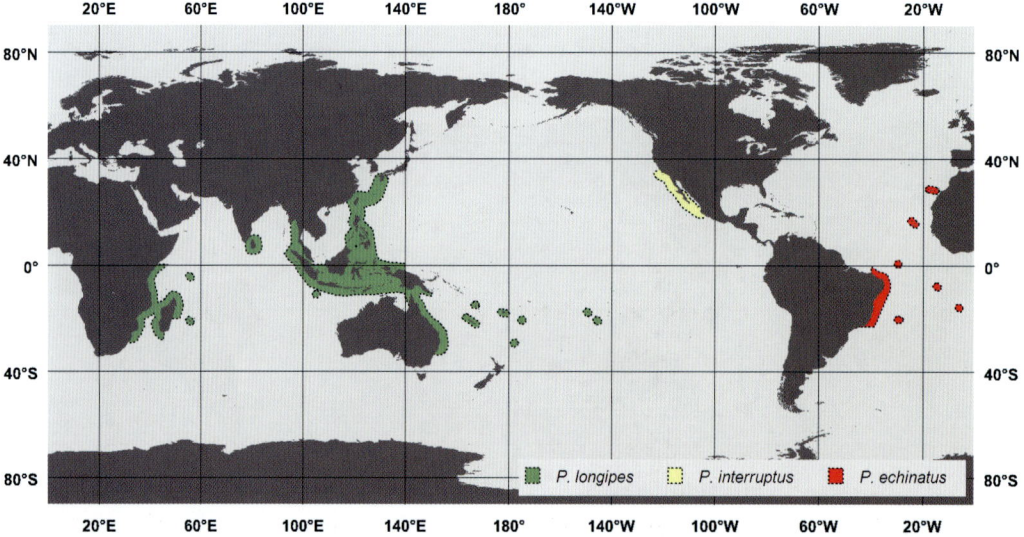

Indicative distribution of the longlegged spiny lobster *Panulirus longipes*, the California spiny lobster *Panulirus interruptus*, and the brown spiny lobster *Panulirus echinatus*.[3,29,37,39–42] (A much wider distribution of the longlegged spiny lobster than that shown has been suggested for the west South Pacific Ocean.[43]) Note, however, that *P. longipes* here is actually a mix of two species, *P. longipes* (as two subspecies) and *P. femoristriga*.[22] The maps cannot be updated to take this into account until the specimens have been re-examined.

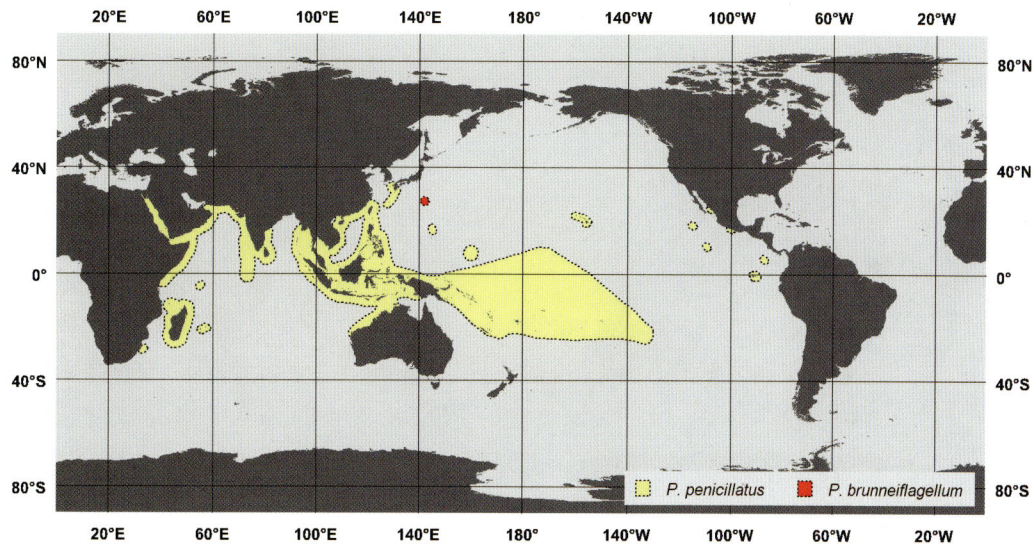

Indicative distribution of the pronghorn spiny lobster *Panulirus penicillatus* and aka-ebi *Panulirus brunneiflagellum*.[3,21,37,44]

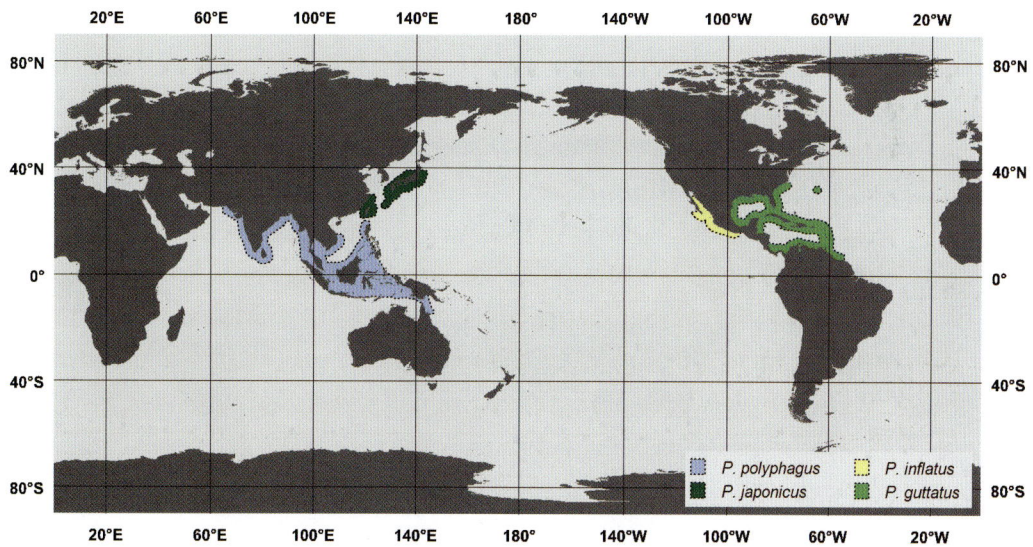

Indicative distribution of the mud spiny lobster *Panulirus polyphagus*, the Japanese spiny lobster *Panulirus japonicus*, the blue spiny lobster *Panulirus inflatus*, and the spotted spiny lobster *Panulirus guttatus*.[3,30]

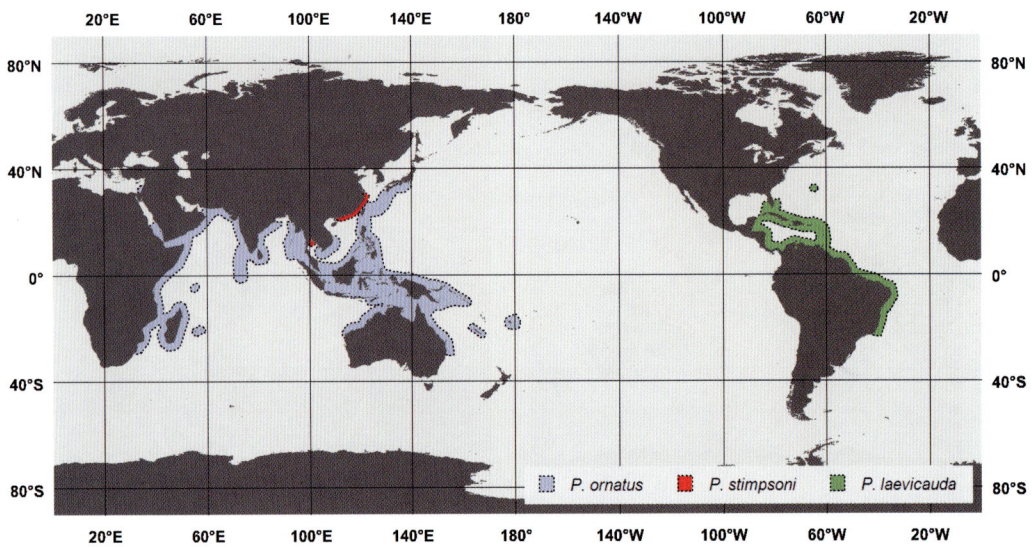

Indicative distribution of the ornate spiny lobster *Panulirus ornatus*, the Chinese spiny lobster *Panulirus stimpsoni*, and the smoothtail spiny lobster *Panulirus laevicauda*.[3,37,45]

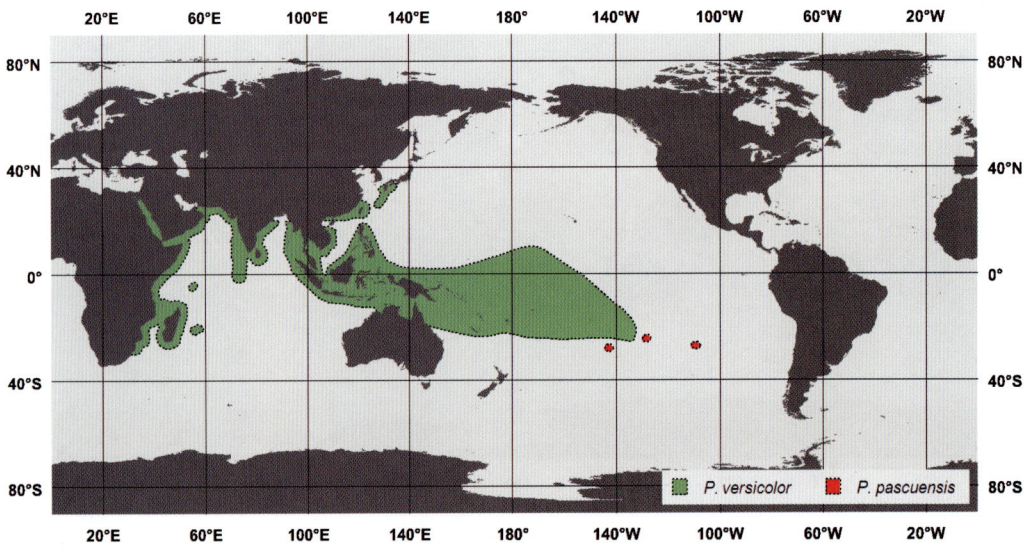

Indicative distribution of the painted spiny lobster *Panulirus versicolor* and the Easter Island spiny lobster *Panulirus pascuensis*.[3,37,46]

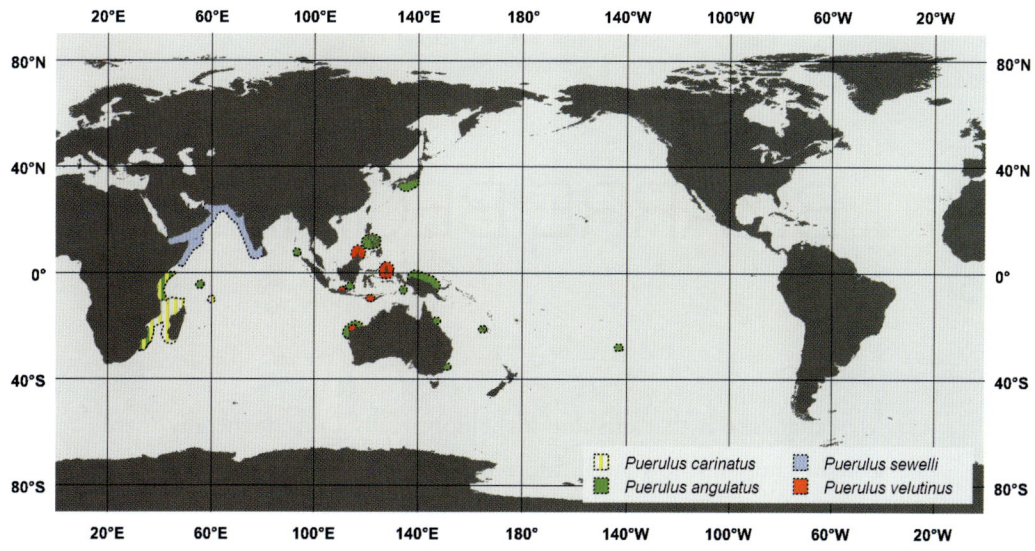

Indicative distribution of the red whip lobster *Puerulus carinatus*, the banded whip lobster *Puerulus angulatus*, the Arabian whip lobster *Puerulus sewelli*, and the velvet whip lobster *Puerulus velutinus*.[3,29,37]

CHAPTER 4

BIZARRE BEGINNINGS

My experience is that offshore it's seldom really calm. Generally there's at least a slight chop and low swell; more often the waves are close to breaking under the persuasion of persistent wind. But this evening the breeze and swell are best described as benign, and there is little of interest to see from the bridge of the 42-metre research trawler *James Cook*. Even the seabirds have vanished as we wait for night to settle so sampling can begin. But the depth sounder shows that there is a lot of action taking place below. On the screen is a steadily upward-moving blanket of animal life emerging from the depths. It gets so dense that the alarm on the sounder warns that the waters are too shallow for this ship—except that here, 80 kilometres off Gisborne on the east coast of the North Island, it is more than 3 kilometres deep.

This is the deep-scattering layer, a wild cacophony of weird form, brilliant colour, and crystal-clear transparency. The smallest items are too tiny to be retained by our net. The largest are up to 20 metres or so, tubes composed of thousands of individual salps into the heart of which a diver could swim. From time to time their broken, slimy remnants contaminate the net.

Most members of the deep-scattering layer are slow-swimming animals—the zooplankton—but among it are the much more mobile denizens such as the fishes—the nekton. The fundamental reason for this invasion from below is the zillions of microscopic plants—phytoplankton—that live in the euphotic zone near the ocean surface, where they are powered by the sun during the day. Rising to feed on the phytoplankton at night, when they are least likely to be seen by predators, are the often transparent, filter-feeding zooplankton. And with these mainly small animals come larger and larger predators. Among

The plankton is sampled with a huge, fine-meshed net (left). This sample contained mainly krill but among it is a host of transparent as well as pink-tinged animals, bizarre-looking fish . . . and spiny lobsters (right).

the more abundant is a wide array of shrimp-like crustaceans, many pink or crimson. Being transparent or dark in colour—mainly reddish—helps them avoid detection. (Red light is the first to be absorbed when light penetrates water, so red objects lit by dim light at depth appear black.) Among the larger of the nekton retained by our net are fish seemingly not of this world: anglerfish with a lighted lure hanging within range of a cavernous mouth; fish that appear as black blobs of jelly but when examined more closely reveal on one edge a mouth, full of long, curved teeth; gulpers, pitch black and eel-like but with a disproportionately large mouth full of equally large teeth; snipe eels, needle-like with delicate pointed jaws, some a metre in length; lightfishes and hatchetfishes, solid of body and with numerous light-emitting panels appearing as mirrors along their bellies and sides. These and others like them are among the top predators in the deep-scattering layer, many having the large mouth and long sharp teeth characteristic of opportunistic pouncers.

It is out here, where waters are very deep, that the leaf-like larvae (phyllosomas) of even the shallow-water spiny lobster species such as packhorse must go if they are later to take their place on the inshore reef. Within a few months of hatching near the coast, the larvae have been transported tens to hundreds of kilometres offshore. At night they come close to the surface to feed and as dawn approaches they depart to the depths to avoid the predators.

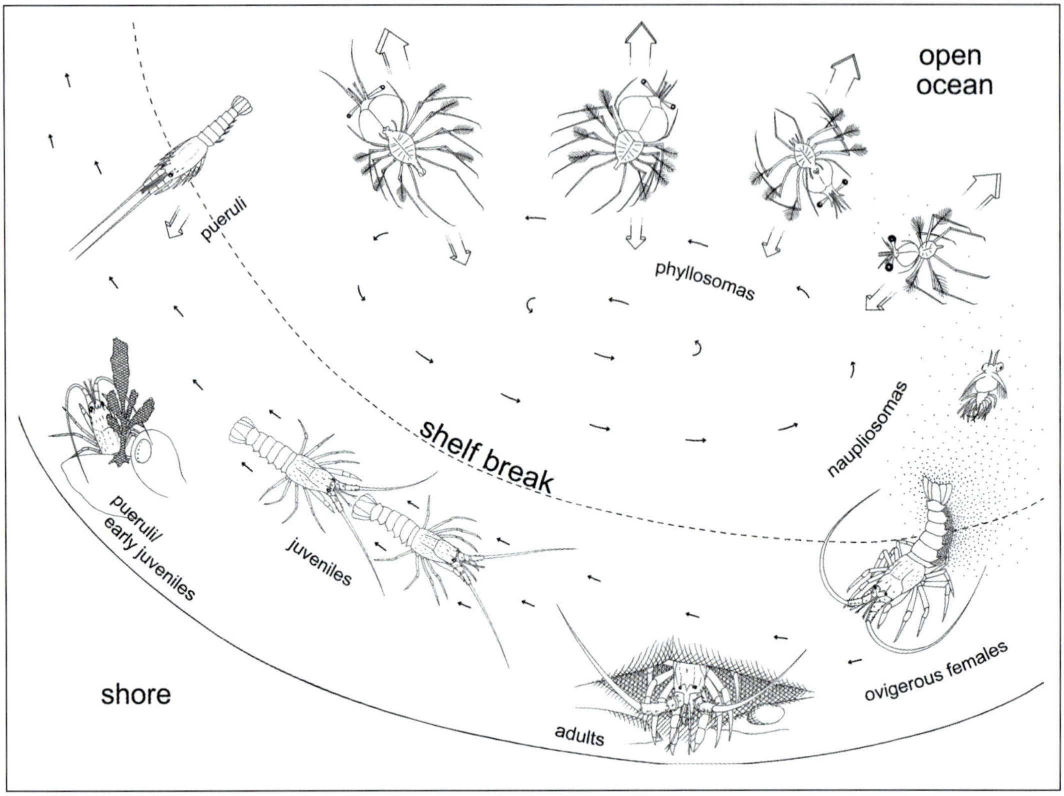

A generalised plan of the life cycle of shallow-water spiny lobsters. (The life cycle of deep-water species involves the same life-history phases, but next to nothing is known of their development or ecology.) Ovigerous females are those bearing external eggs.

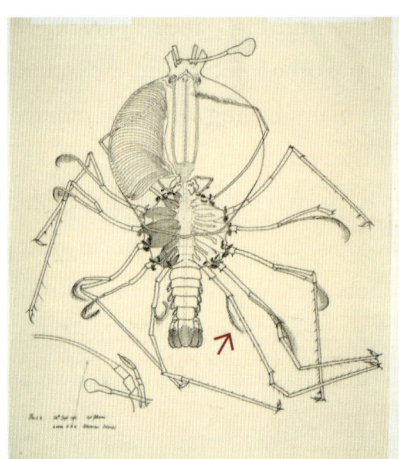 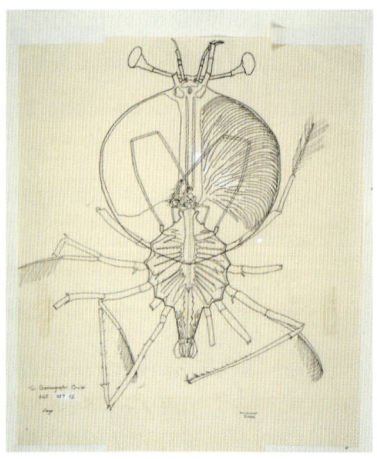

Richard Pike's drawings of the final-stage phyllosoma of the packhorse (left, a view from above) and of the red rock lobster (right, from below). Each larva is about 40 mm long from between the eyes to the tip of the tail. The packhorse phyllosoma has a less rounded body, and an extra exopod on its fifth legs (see arrow).

In a large-format, dusty drawing book, covered in a pink floral pattern, is the first illustration of a late-stage phyllosoma larva of the packhorse, from almost 50 years ago. This larva had been taken in a plankton tow in the Bay of Plenty, on the east coast of New Zealand's North Island, in September 1962. It, together with phyllosomas of other species, had been drawn in pencil soon after by Richard Pike, the Marine Department's first crustacean scientist. It's unclear just who recognised the significance of Richard's illustration, correctly labelling it in pencil *verreauxi* in contrast to others, very similar-looking, identified as *edwardsii*—the handwriting is not Richard's. But whoever it was had unmistakably identified the key feature that distinguishes the phyllosomas of these two species. In addition to being somewhat narrower and more like an inverted pear, Richard's final-stage packhorse phyllosoma shows an extra branch on one of its appendages. On each of the last pair of legs is a hairy swimming appendage (exopod) which is absent on the red rock lobster phyllosoma. It would be a further quarter century before this crucial difference was published, based on plankton collections off the east coast of Australia.[1]

Packhorse have the most eggs per clutch of any spiny lobster: a female, depending on size, will lay between half a million and 2 million or so.[2] They develop within the protection of the brood chamber formed by their mother's tightly curled tail. Under a dissecting microscope the eggs, each just 0.6 millimetres in diameter, appear first as bunches of brick-red grapes. They change colour as they develop, first becoming amber and then muddy brown when seen en masse. Just before hatching the large black eyes of the developing larvae, and the flashes of red pigment on their tightly curled legs, become visible through the eggs' thin membranes.

Packhorse eggs hatch in early summer into a short-lived, 1.5-millimetre-long naupliosoma larva, a cartoonlike creature with huge eyes and a compact body. This phase seems to be a feature of only the more primitive spiny lobster species, a hangover from ancient times. Within a matter of minutes or a few hours, the naupliosoma moults into the gangly-legged first-instar phyllosoma larva. The phyllosoma then moults more than a dozen times over the following months, each instar (developmental form between moults) being a little larger and more complex in form. (The many instars are arbitrarily grouped into stages of development for simplicity; there is often more than one instar per stage.)

During this time the phyllosoma is transported into the open ocean, well out of sight of land. After 9 months or so, it reaches its final instar. Now close to 40 millimetres long, it is more than an order of magnitude larger than it was when it became a phyllosoma, but it still doesn't resemble a spiny lobster. It looks more like a large spider, squashed—a flat, crystal-clear, leggy creature of very strange form indeed.

Phyllosomas at all stages of development have the same body plan, but the older they are the greater their number of appendages. There are three main parts to the body. First is the cephalic shield with its paired antennae (feelers) and eyes at the front, and its paired maxillipeds (feeding appendages) and mouthparts underneath. Then there is the thorax, from which emerge five pairs of long, segmented pereopods (swimming legs). Last is the abdomen (tail), with its pleopods (swimmerets) and its tail fan, as identifiable in late-instar phyllosomas as the tail of the adult. Rick Webber's illustration later in this chapter of a final-stage packhorse phyllosoma clearly shows the distinguishing exopods on the last pair of legs.[3] It also displays a further key feature that distinguishes the packhorse phyllosoma from that of the red rock lobster: an exopod on the largest of the feeding appendages, the third maxillipeds. (The presence of this exopod, however, seems to be intermittent—it was not, for example, present on Richard Pike's drawing of a packhorse phyllosoma.) The exopod on the fifth leg first

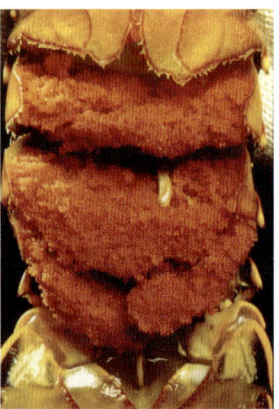  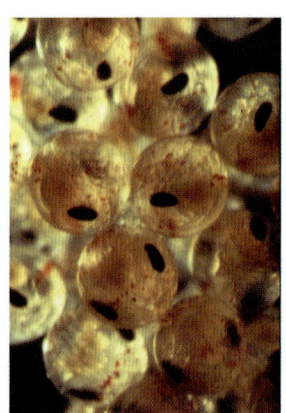 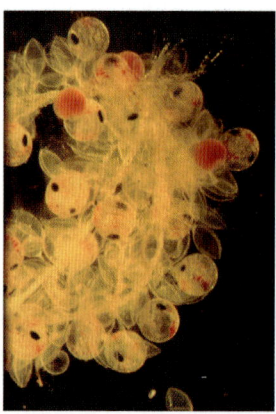

Packhorse have the largest egg clutches of all spiny lobsters (far left). Recently extruded eggs (middle left) show little sign of the developing embryo, but close to hatching the paired eyes and red pigment spots on the legs are obvious (middle right). Ruptured egg capsules in an egg mass at hatching (far right) confirm that some of the naupliosomas (newly hatched larvae) have flown.

becomes evident early in the larva's development whereas the exopod on the third maxilliped, when present, first appears halfway through the larva's development, and is not fully formed until the final instar.

At the end of the phyllosoma phase, a profound and dramatic change in body form takes place. From being large and leaf-like, the lobster is transformed into the smaller, more cylindrical form of the next phase, the postlarva, or puerulus. The cephalic shield and thorax of the phyllosoma become fused into the cephalothorax of the puerulus. Now for the first time the lobster resembles the adult, although it's still largely transparent.

Few packhorse phyllosomas have been captured in the wild. No packhorse puerulus has ever been reported caught in the plankton, although they have been occasionally found on New Zealand shores and they are common during spring on collectors set along the New South Wales coast.[4] But this does not mean that we are bereft of information on the form and functioning of either life-history phase, as the full development of packhorse is well known through larval culture. In northern Japan in 1990, Jiro Kittaka was the first to successfully culture packhorse larvae, growing close to 200 from egg to puerulus.[3] Taking between 6 and 12 months to develop to the final instar, they moulted 17 times. Many more packhorse phyllosomas have been cultured since, in both New Zealand and Australia.

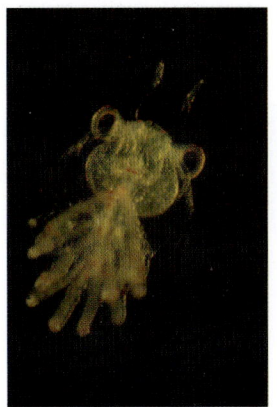 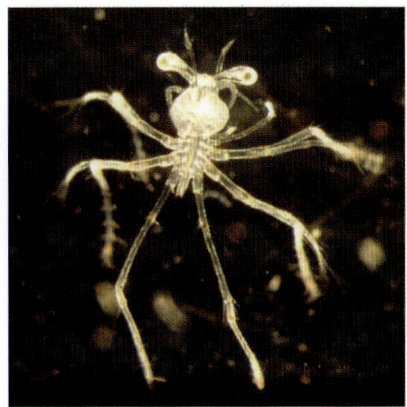 

The packhorse egg hatches into a 1.5-millimetre-long naupliosoma (left) that within hours moults into a first-instar phyllosoma (middle). After many months and many moults, the phyllosoma has grown to almost 30 times its initial size (right).

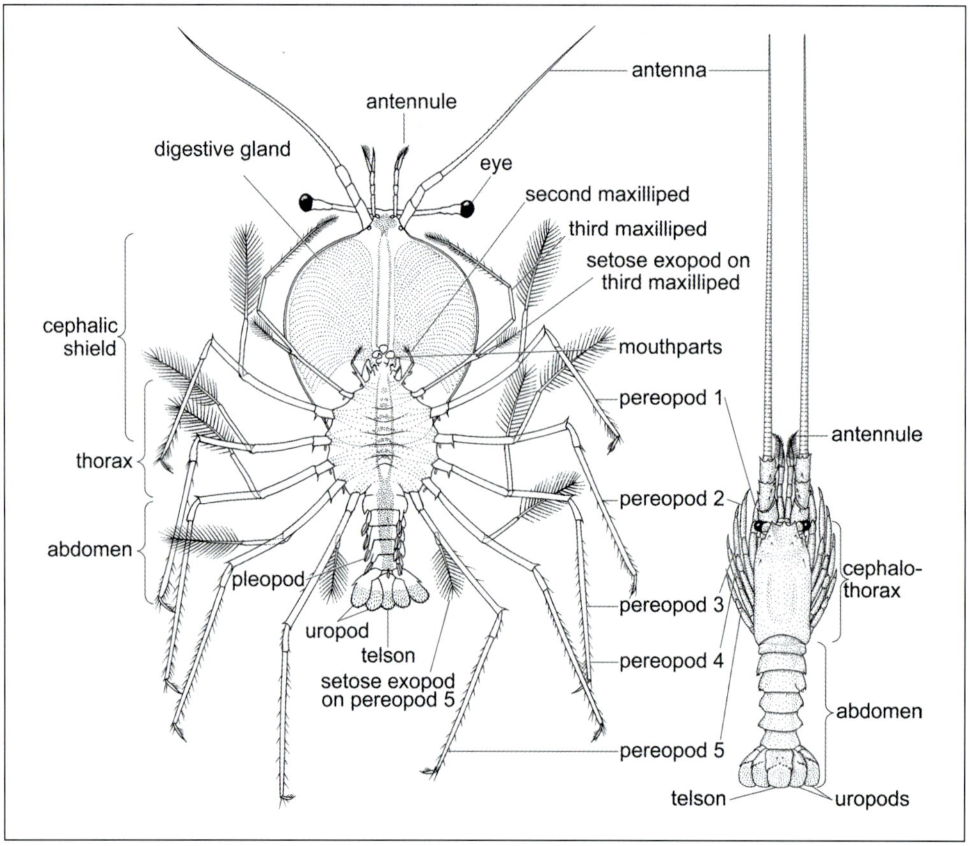

Body parts of the final-instar packhorse phyllosoma (left, from below) and the puerulus into which it metamorphoses (right, from above). The phyllosoma is 40 millimetres long from between the eyes to the tip of the tail, and both are drawn to the same scale.

The final-stage phyllosoma of both the packhorse and the red rock lobster would cover a good portion of your palm (right). The puerulus immediately after metamorphosis is smaller and essentially transparent—and well equipped to steady itself by grasping (far right). But within a few days the reddish-brown shell of the juvenile packhorse becomes visible through the transparent shell of the puerulus. Older juveniles are dark green.

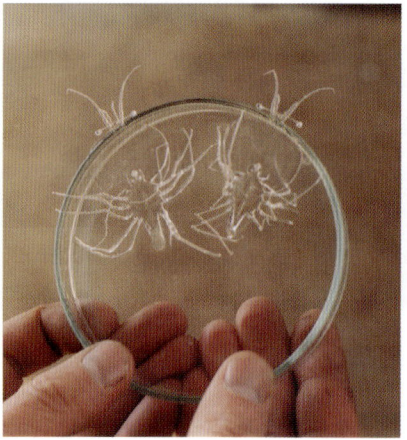
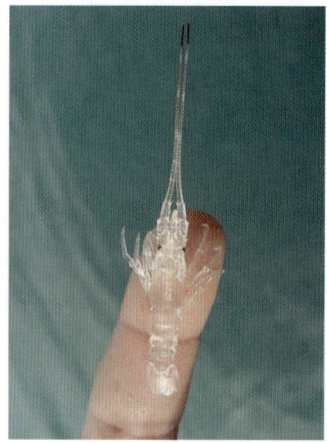

Culturing packhorse phyllosomas in the laboratory has allowed insights into the behaviour of this bizarre early life-history form. They swim not in straight lines, but by looping and tumbling.[15] Their mouthparts, used to grind and chew as they are in the adult, are tucked seemingly inconveniently half way back along the underside of the cephalic shield. These rather ludicrous-looking creatures are nevertheless effective killers, surprisingly efficient at detecting and pouncing on their food. Newly hatched brine shrimp larvae have most often been provided as food for the first few instars, and small cubes of mussel gonad or larger brine shrimp for the later ones. Serena Cox, at the National Institute of Water and Atmospheric Research (NIWA), scrutinised the feeding strategies and behaviour of this rapacious predator.[16]

> Once impaled by a dactyl (the spear-like tip of a leg), the defenceless prey is torn into smaller pieces, limb by limb, ripped apart in a somewhat frenzied feeding attack, and shoved remorselessly between the mandibles to be ground, pulped and devoured. The entire attack may take only a matter of minutes, depending on the fight put up by the prey.
>
> The phyllosomas' six pairs of mouthparts provide a veritable array of cutting, shearing, stabbing and grabbing utensils, capable of penetrating the hardened exoskeletons of krill yet delicate enough to extract the soft, juicy gut of a copepod. The sharp dactyls are perfectly formed for piercing and grasping any unsuspecting prey swimming by. Closer to the main part of their body, the maxillipeds are covered in robust and elongate setae (long hairs) which, with their razor sharp tips and serrate edges, firmly adhere to the prey. During development, these setae become more robust and numerous, allowing the phyllosomas to grasp and devour larger and larger prey.
>
> The maxillae help push prey items towards the mandibles, at the same time tearing and shredding. The second maxillae also trap small food particles, pushing and shoving these morsels of food lost during the ferocious shredding activity to the mouth. Tucked away behind the maxillipeds and maxillae are the hidden weapons, the armoured, calcified and serrated grinding plates of the mandibles. The sharp incisor and molar processes cut food into smaller pieces, and then grind it into particles small enough to be sucked through the small mouth into the foregut and thence into the simple digestive tract.
>
> Like any efficient predator, cleaning up after a kill is equally as important. The maxillipeds meticulously pick and scrape food scraps

IDENTIFYING FINAL-STAGE PHYLLOSOMAS

When first reported, phyllosomas were thought to be distinct animals in their own right belonging to the genus Phyllosoma.[5] *We now know better. But the phyllosoma larva you have just fished out of the net, or found preserved in some antediluvian collection, could be any one of close to 140 species. First, how do you distinguish the phyllosomas of spiny lobsters from those of slipper lobsters? OK, so it is a spiny lobster—but what genus? The genus (but seldom the species) of a final-stage (and, in most cases, penultimate-stage) larva can be determined from the length ratios of its body parts and how well developed its appendages are. (Final-stage larvae have well-developed gills on the bases of their pereopods.) Here Rick Webber and I show how to be sure that the late- or final-stage phyllosoma is a spiny lobster, and how to identify it to genus. Note that* Justitia *cannot be distinguished from* Nupalirus *until more of these larvae of very similar appearance have been described. Also, the form of the final-stage phyllosoma of* Linuparus, *and all larval stages of* Projasus *and* Palibythus, *are as yet unknown, although* Linuparus *larvae appear to be reliably identified by the presence of a projection on the outer sides of their antennae.*

1. Distinguishing phyllosomas of spiny lobsters (family Palinuridae) from those of slipper lobsters (family Scyllaridae) Just as in the adults, spiny lobster larvae have long, whip-like antennae whereas in the slipper lobsters they are short and flat (see facing page).

2. Distinguishing the phyllosomas of spiny lobster genera Outline illustrations of the form of the final-stage phyllosomas of genera of the Palinuridae appear opposite. The larvae are not drawn to scale: *Jasus*[6] and *Sagmariasus*[3] phyllosomas typically reach 40–50 millimetres total length (the distance from between the eyes to the tip of the tail); *Panulirus*[7] and *Justitia/Nupalirus*[8] 35 millimetres; *Palinurus*[9] 20 millimetres; *Palinustus*[10] 50 millimetres; and *Palinurellus*[11] and *Puerulus*[12] 25 millimetres. Legs of only one side are illustrated. The phyllosomas of *Palinurellus* and *Puerulus*, unusual in that the cephalic shield completely (or nearly so) covers the thorax, have sometimes been referred to as 'phyllamphion' larvae.[12,13]

Key to identify to genus the final-stage phyllosomas of spiny lobsters

1	(a)	Antenna long and whip-like	2 (Palinuridae)
	(b)	Antenna short and flat	Scyllaridae
2	(a)	Cephalic shield covers, or almost covers, the thorax	3
	(b)	Cephalic shield covers only part of the thorax	4
3	(a)	Cephalic shield almost circular	*Puerulus*
	(b)	Cephalic shield almost rectangular, with a pointed rostrum	*Palinurellus*
4	(a)	All pereopods about equal in length	5
	(b)	5th pereopod smaller and shorter than the rest	6
5	(a)	5th pereopod with an exopod	*Sagmariasus*
	(b)	5th pereopod without an exopod	*Jasus*

6 (a)	Cephalic shield no wider than the thorax		7
(b)	Cephalic shield wider than the thorax		8
7 (a)	At least pereopods 1 and 3 chelate or subchelate		*Justitia/Nupalirus*
(b)	No pereopods chelate or subchelate		*Panulirus*
8 (a)	Pereopods 1–4 similar in length and chelate or subchelate		*Palinustus*
(b)	Pereopods 1–3 long, the second the longest; none chelate or subchelate		*Palinurus*

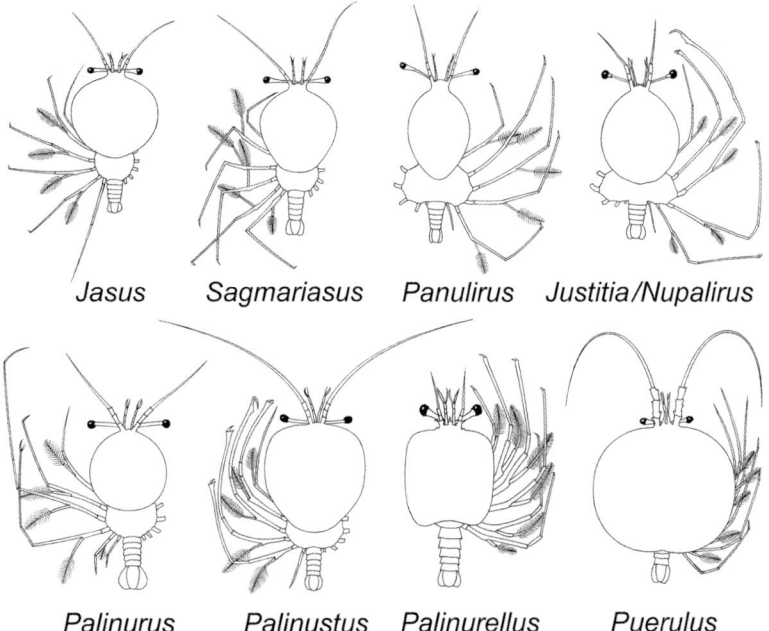

Jasus   *Sagmariasus*   *Panulirus*   *Justitia/Nupalirus*

Palinurus   *Palinustus*   *Palinurellus*   *Puerulus*

3. Distinguishing the phyllosomas of New Zealand's slipper lobsters Below are outline illustrations of the final-stage phyllosomas of scyllarid species known so far from New Zealand waters (the last being rare).[14] *Ibacus alticrenatus* phyllosomas reach 40 millimetres total length; *Antipodarctus aoteanus* 30 millimetres; *Arctides antipodarum* 55 millimetres; and *Scyllarides haanii* 50 millimetres.

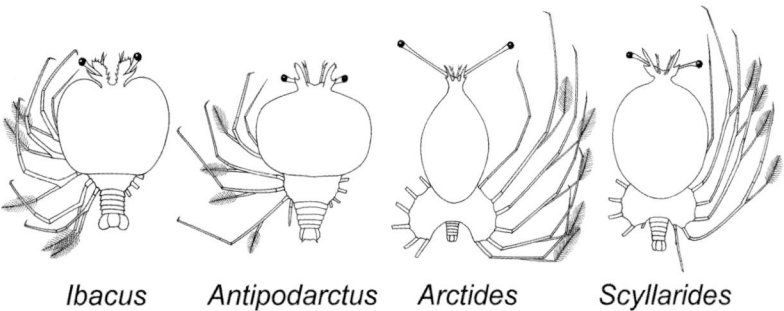

Ibacus   *Antipodarctus*   *Arctides*   *Scyllarides*

from the mouthparts and surfaces of the phyllosoma. Broad sweeping movements comb the setae and spines of the maxillae, scraping off remaining fragments of a now unidentifiable former member of the zooplankton.

So the mouthparts of packhorse phyllosomas—like those of other spiny lobsters—are well adapted for grasping, manipulating, cutting and masticating foods. Improved processing efficiency and prey handling come with development, prey size also increasing. Under culture conditions, early-stage phyllosomas are probably best fed a small particle diet with soft texture (such as a gelatin/alginate base) whereas later instars are capable of consuming a diet with larger particles and a firmer, fleshier consistency. By late stage the phyllosomas are perfectly capable of tearing and shredding even large, firm and fleshy prey.

NIWA's Len Tong observed how packhorse phyllosomas 'fuelled up'. Nailing several food items in quick succession, each on a different leg tip, they would then proceed at leisure to transfer them one-by-one to their mouths for chewing. This would be handy in nature, the phyllosomas being able to withdraw sooner to the safety of the dark depths where they can masticate and consume their food in relative peace.

Jiro Kittaka revealed how metamorphosis in packhorse begins with the new shell steadily separating from the old.[3] The phyllosoma stops feeding and its digestive gland—similar in function to the mammalian liver—shrivels. On the big coming-out day, the phyllosoma is less keen on swimming, instead constantly flexing its abdomen and beating its pleopods vigorously. The newly formed cephalothorax emerges from the old cephalic shield and thorax and then, as the old shell is bent back and forth, the abdomen appears, followed by the pereopods. Finally the antennae emerge as the puerulus flexes against the water current. The whole business takes just 20 minutes.

These new pueruli are equally as well camouflaged for their time in the plankton as their antecedents. They are crystal-clear, except for their black eyes and the dark bands towards the tips of their antennae, and the somewhat more ephemeral red pigment spots often visible on the legs. But the puerulus is a stronger swimmer than the phyllosoma. It advances with its antennae and pereopods extended ahead, using its relatively large pleopods for propulsion—just like a shrimp in a rock pool. When disturbed, it retreats in a series of rapid

PICKING A PUERULUS

You have found a lobster-like animal among shallow rocks and seaweed. It is essentially transparent, and it has no prominent nippers. It keeps bending its tail up under its body and then releasing it with a jerk in its bid to escape. But is it a puerulus?

If you saw it active out in the open during the day, it is unlikely to be a puerulus: unless disturbed, pueruli are strictly night-workers. With its tail outstretched, it will be no less than 12 millimetres long from between the two antennae to the tip of its tail. Most pueruli are between 15 and 25 millimetres in total length, although that of *Projasus*—which yours is most unlikely to be—is a gigantic 50 millimetres. It will be flat from top to bottom, whereas shrimps—with which pueruli are most often confused—are flattened laterally. The tail will be of similar width to, and similar length or longer than, the body. In the megalopa, the life-history stage of crabs equivalent to the puerulus, the short tail is tucked up under a comparatively bulky carapace. If the antennae are long and whip-like, then your find is likely to be a spiny lobster puerulus; if they are plate-like, then it's probably a nisto—the slipper lobsters' equivalent to the puerulus.

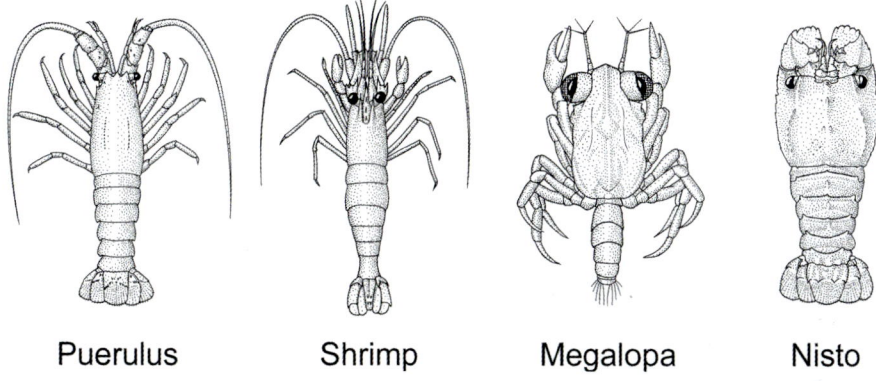

Puerulus   Shrimp   Megalopa   Nisto

The transparency thing is a little complicated. As we have learnt, a puerulus starts off being virtually transparent, but as it approaches the moult it darkens. Tease out its pleopods. If they are so small that they are barely visible with the naked eye, you are probably clutching a first-instar juvenile spiny lobster; if they are so long that they overlap, it's a puerulus.

Too few pueruli have been described to allow a useful key to be constructed. But it is possible to distinguish the puerulus of the packhorse from that of the red spiny lobster, although you'll need an eye-piece. The main difference is that, just as in the adult, the rostrum (the forward-projecting middle part of the top front of the carapace) of the packhorse puerulus is flat and about the same size as the supraorbital spines that protect each eye; in the red spiny lobster, the rostrum is much smaller than the supraorbital spines.

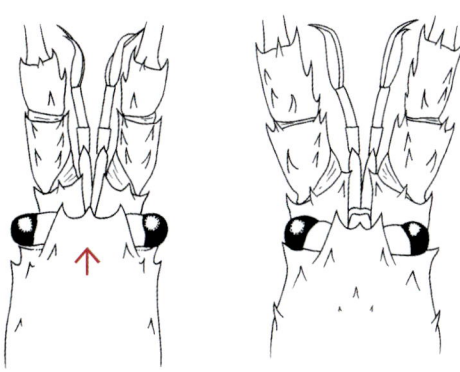

Puerulus stage of the packhorse (left), and the red rock lobster (right).

backward tail-flicks. The primary fuel of this non-feeder is a class of fat known as polar lipid, which is transparent.[17] It took 3 weeks before Jiro's pueruli were ready to moult into juveniles. The digestive gland had become visible a week earlier, and pigmentation had gradually extended over the entire body. The juvenile that emerged had all the characteristics of the adult.

φφφ

Learning about water flows in the Far North likely to affect the distribution of phyllosomas was crucial in understanding the recruitment mechanism (how new individuals are added to the breeding population) for packhorse in New Zealand waters. After all, 9 months or so in the plankton is potentially time enough for a poor swimmer to be transported great distances by the ocean currents.

Te Rerenga Wairua—more widely known as Cape Reinga—is of enormous significance to Māori. The solitary pohutukawa clinging precariously to the eastern side of the jagged rock promontory marks the 'place of leaping', where the spirits of the dead plunge into the sea before making their way to their final resting place in the ancestral homeland of Hawaiki. Sweeping southwest is the dull-yellow dune belt of Te Werahi Beach, its sands driven inland a kilometre or so by storms. At the western tip of the beach is Cape Maria van Diemen, named in 1643 to honour the wife of the Governor of Batavia by Dutch explorer Abel Tasman, the first known European to visit New Zealand. To the east the immediate view is scrub-covered hills, interrupted by red soil slips and tracks, with the dunes of Spirits Bay and the headland of the Surville Cliffs in the distance. On the horizon to the north are the dark shadows of the Three Kings Islands. Below Cape Reinga is the frothy turbulence of swells pitching onto

A lone pohutukawa tree clinging to the eastern side of the extreme promontory of Cape Reinga marks the 'place of leaping'.

the shallow Columbia Bank in all but the calmest conditions. This is sometimes incorrectly said to be the result of the Pacific and the Tasman 'colliding'.

The Far North has almost certainly been New Zealand's most important packhorse breeding area for eons—at least since the islands of the north were linked by the sand tombolo to the mainland about quarter of a million years ago.[18] The waters here contain remarkable seafloor biodiversity. There is an assortment of sponges and other invertebrates not known anywhere else, an unmatched richness of bryozoans (moss animals).[19] The substrate itself is made up of the calcareous remains of these organisms together with those that once lived in the waters above, with little land-sourced sediment at all. This is also an area of strong tidal mixing and upwelling,[20] creating cool, nutrient-rich surface waters, with accompanying lush phytoplankton blooms that are particularly pronounced during summer.

After the female packhorse have extruded their eggs inshore in October, both the males and females make their way across the bryozoan and sponge fields into deeper waters, where egg-hatching takes place during December and January. Presumably the invertebrate commons provide plenty of food on the way, and the rich waters and strong currents assist in both the sustenance and dispersal of the young larvae.

When studies began in the 1970s into the recruitment mechanism for packhorse, the flows likely to play a role in the dispersion of the weak-swimming packhorse larvae were not at all well known, and the northward migration of juvenile packhorse up the east coast of the North Island was yet to be discovered. The little information available on phyllosoma ecology and distribution was for species much more widespread than packhorse. These included the red rock lobster, which breeds throughout all of mainland New Zealand and southern Australia; the Cape rock lobster *Jasus lalandii*, which breeds throughout its 1000-kilometre distribution along the west coast of southern Africa; and the western rock lobster *Panulirus cygnus*, which breeds throughout its 1200-kilometre range along the shores of Western Australia. The larvae of each of these species had been found to develop in waters well offshore. But with New Zealand packhorse breeding in essentially just one small area in the far north of the country—a region of overall west-to-east water flow to boot—surely any larvae carried too far from shore would be lost altogether from the region. How were phyllosomas being retained nearshore?

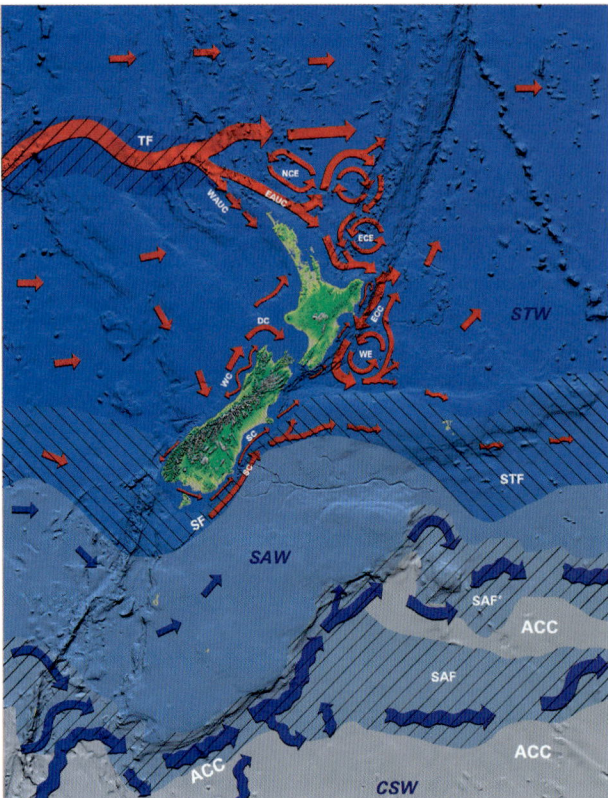

Surface currents assail New Zealand from the west. The Tasman Front (TF) flows directly from Australia. Two currents derived from it—the East Auckland Current (EAUC) and the less persistent West Auckland Current (WAUC)—affect the coasts of Northland. The inshore flow south of East Cape is known as the East Cape Current (ECC). The North Cape Eddy (NCE), the East Cape Eddy (ECE), and the Wairarapa Eddy (WE) also greatly influence the final destination of packhorse larvae. Other currents are the D'Urville Current (DC), Southland Current (SC), and the Westland Current (WC). (SAW=Subantarctic Water; SF=Southland Front; STF=Subtropical Front; STW=Subtropical Water; ACC=Antarctic Circumpolar Current; SAF=Subantarctic Front; CSW=Circumpolar Surface Water.)

Two mechanisms were identified that potentially retain packhorse larvae nearshore for many months. First, back eddies associated with points jutting out into the coastal current were well known elsewhere to retain fish and other larvae;[21] Great Exhibition Bay, in the lee of North Cape which pokes out into the south-flowing East Auckland Current, was one such candidate location. Second, the intermittent presence of the West Auckland Current—a weak, south-moving flow on the west coast of Northland, sometimes reaching as far south as Manukau Harbour and detectable up to 100 kilometres from shore[22]—could help to retain larvae near the coast. The primary tools used to test if either mechanism played any role in keeping packhorse larvae near the breeding area were drogues—weighted screens (paravanes) suspended beneath surface floats.

The peaks of land that give texture to the otherwise low-lying tombolo of the Far North lent themselves to shore-based tracking of radio-transmitting drogues. Four drogues, their paravanes suspended either 25 or 50 metres below the buoy, were launched between Cape Reinga and the Three Kings Islands in 1978, at the height of the packhorse egg-hatching season. Each transmitted a unique signal, detectable from any high point within a direct line of sight. All four drogues went west and south, one stranding just north of the Kaipara Harbour, which is entirely consistent with the presence from time to time of the enigmatic West Auckland Current.

Next we tried radar-tracked drogues, which proffered different prospects— and problems. It was not going to be possible to readily update their positions. On the other hand, their usefulness was not going to hinge on the longevity of their transmitting battery or whether they remained within the line of sight. The reflectors on the drogues were proven detectable by the radars of the Royal New Zealand Air Force Orions; the colour of the reflector allowed individual drogues to be identified. Seven radar-reflecting drogues, their paravanes at a depth of either 30 or 60 metres, were set along a line between Cape Reinga and the Three Kings Islands in 1979, four in April and the remainder in December. The positions of these drogues were to be determined during fishery surveillance flights by the Orions.

Not one of the seven radar-reflecting drogues was ever detected by the Orions, but the destination of five of them is known. Two arrived near the entrance to Kaipara Harbour a couple of weeks after deployment. One beached in the northern part of Great Exhibition Bay after 3 weeks, and another at the northern end of Ninety Mile Beach after 4 weeks. And the fifth drogue travelled considerably further.

In mid-November 1981, Fraser MacLean, Fisheries Officer at Kaitaia, received a letter from C.J. Mee, Manager of Milne Bay Islands Fisheries, Misima Island, in the heavy but uneven font of a manual typewriter.

> . . . A buoy with the figure 1 and your address was brought into my office this morning. It was washed ashore on the beach at Siagara Village on the North Coast of Misima Island at a position Lat 10° 38' South Long 152° 46' East. The local people watched it being washed ashore time approx 1500 hours, 26 Oct 1981 . . .

Misima Island is off the northern shore of the southeast tip of Papua New Guinea, thousands of kilometres from Cape Reinga. Who knows what the radar reflector, standing almost a metre above the buoy, had witnessed during the 2.5

years between its launch on 24 April 1979 and its arrival in Papua New Guinea on 26 October 1981. How many times was the reflector beaten to the surface of the sea by spindrift, even entirely submerged? Was it ever glimpsed on the radar screen of a ship or aircraft—but not investigated because there was no time or opportunity? Subsequent correspondence from Chris Mee revealed that the drogue was, unremarkably, missing its paravane, the pattern of fading on the buoy indicating it had broken away some time before. Nevertheless, with the radar reflector itself then acting as a drogue, its drift would still not have been solely the result of wind acting on the extreme sea surface—as it is with a floating bottle. Chris Mee later informed us that the remaining metal componentry of the drogue was unlikely to be returned: 'I've no doubt it is already in use as fish spears or pig bashers or the like.' (No better an outcome, I'd say.)

Probably this drogue, its paravane suspended 30 metres down, first went east but also north, to be caught up in the flow associated with the west-directed belt of the southeast trade winds which lies closest to northern New Zealand during summer. But did the drogue complete an entire circulation of the South Pacific Ocean, being seen close to the shores of South America before being carried north, then west? All we know is that the *minimum* distance it drifted was 3500 kilometres, at a minimum rate of almost 4 kilometres each day.[23]

Vessel-tracked drogues were the other tool used to track the currents, making use of the disproportionate significance assigned to floating objects glimpsed at sea. The buoy supporting each paravane carried instructions for the finder to promptly return the drogue to the sea and to immediately report the position. In 1978, ten of these were launched: three off Cape Reinga on a line towards the Three Kings Islands, three on a line east from Great Exhibition Bay, and four along a line west from Ninety Mile Beach. Although the position of only four of these drogues was ever reported, they showed remarkably similar journeys to the radar drogues. One, set 8 kilometres off Cape Reinga, was reported 43 days later off Cape Brett; another, from Ninety Mile Beach, came ashore almost due east. But interestingly the buoy of one of the vessel-tracked drogues, launched 30 kilometres northwest of Cape Reinga with its paravane set at 30 metres, had—almost unbelievably—also reached Papua New Guinea, this one within 4.5 years of its deployment. The following telegram was received on 11 May 1983 from the New Zealand High Commission in Port Moresby:

PNG FISHERIES OFFICER AT TUFI HAS SENT US A RED PLASTIC MARKER BUOY WHICH IS MARKED MAF FISHERIES RESEARCH DIV. WITH REQUEST FINDER NOTIFY FISHERIES OFFICE, MAF AT KAITAIA. THE MARKER WAS FOUND AT BLACK ROCKY POINT, NORTHERN PROVINCE OF PNG AT 1745 21 APRIL.

Northern Province is immediately northwest of Milne Bay Province. Remarkably, two drogues, each launched at about the same spot but at very different times, came ashore at points a few dozen kilometres apart after years and thousands of kilometres at sea. This drogue's minimum rate of drift was a little over 2 kilometres each day.

What did the drift of these drogues, together with concurrent sampling of temperature, salinity and the plankton, reveal about larval retention mechanisms—such as eddying or reversals—in the Far North? Well, they provided no evidence for any long-term eddies. But they did suggest variability in the nearshore flow, low rates of transport, and the existence from time to time of the West Auckland Current. Each of these physical attributes of the seas close to Northland could potentially assist in retaining packhorse phyllosomas—if the phyllosomas did not go great distances out to sea.

Later, however, as more information trickled in from overseas, it became clear that the larvae of all coastal spiny lobster species—even those such as packhorse with restricted breeding areas—are transported quite rapidly well offshore.[24] For packhorse, therefore, the mechanism of larval retention is not to be found in the lee of North Cape or other promontories, nor in the ebb and flow of the West Auckland Current. Rather, key physical processes taking place well out to sea, including those associated with the more-or-less permanent eddies embedded within the flow of the East Auckland Current, north of both North Cape and East Cape,[25] restrict the extent of dispersal of at least some larvae.

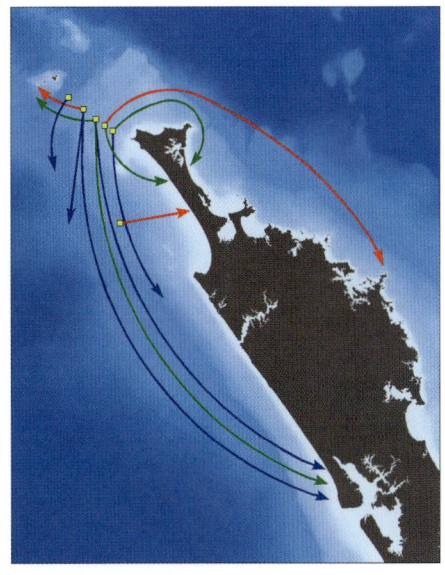

Inferred drift of drogues (blue – radio transmitting; green – radar tracked; red – vessel tracked) deployed in the Far North, 1978–79. The two drogues that reached Papua New Guinea are shown as arrows to the northwest, but there is every chance that each first went north and east before being transported west by the southeast trade winds.

Computer models of ocean currents can be used to simulate where phyllosomas drift after hatching. By adding biological 'behaviour' to the numerical phyllosomas these simulations can be used to model where metamorphosis to the puerulus phase takes place and hence where we expect pueruli to settle. NIWA's Steve Chiswell 'seeded' the Australian Government's Bluelink ocean circulation model with numerical phyllosomas from hatchings in the Far North and calculated the tracks the larvae took as they drifted around the model ocean to their point of metamorphosis and settlement using real-time currents.

By seeding the ocean with many larvae over several years many tracks were generated, which were used to make a statistical estimate of probable larval dispersal. In each simulation year (1995–2005), the model was seeded in summer with a source of newly hatched larvae in the Far North. Trajectories of the larvae were computed through to the end of each settlement window (1 October to 31 December).

Once the track of any individual phyllosoma had been simulated, an algorithm of biological behaviour was used to decide when and where (or even if) that phyllosoma metamorphosed. The algorithm that allowed phyllosomas to metamorphose had three criteria, each designed to mimic what we believe happens in nature. First, metamorphosis could occur only after September of each year because phyllosomas had to be at least 9 months old to metamorphose. Secondly, only phyllosomas within a pre-specified distance of the coast (50, 100 or 150 kilometres) were allowed to metamorphose because metamorphosis in shallow-water species such as the packhorse takes place mainly near the shelf break. Thirdly, the numerical phyllosomas had to have acquired sufficient 'energy' to allow them to metamorphose. This was to account for the pueruli being essentially non-feeding and therefore requiring sufficient energy reserves in the form of lipids to allow them to swim to the coast. The numerical energy level of a phyllosoma was estimated by using chlorophyll levels derived from the SeaWiFS (sea colour) satellite sensor, which are considered to be a good proxy for phyllosoma prey.

During the model runs most larvae were transported great distances off and along the east coast of the North Island, many being lost to New Zealand waters altogether. Nevertheless, some larvae met all criteria. Metamorphosis took place along the entire east coast of the North Island as far south as Cook Strait, together with a small proportion along the west coast of the far north of the North Island (presumably through the influence of the intermittent West Auckland Current). The result was the same irrespective of the specified distance offshore of metamorphosis and required energy levels, suggesting a robust conclusion.

Map of northern New Zealand showing the computed positions of packhorse larvae hatched in summer in the Far North and then transported in the prevailing currents and eddies for 9 months (yellow dots), and the points of metamorphosis for the lucky ones, using an offshore metamorphosis distance of 100 kilometres (red dots).

We can therefore conclude that metamorphosis takes place not far offshore all along the north and east coasts of the North Island. This, as we have already learnt, is where all packhorse phyllosomas and pueruli have come from and where small juveniles have most frequently been reported. A small proportion of larvae appear to be transported south along the west coast, consistent with the presence from time to time of the West Auckland Current. It is not surprising that no very small packhorse have been reported there because much of this coast is sand, no place for a young packhorse to live.

Are packhorse larvae also transported across the north Tasman Sea, from hatchings in New South Wales, as suggested in 1967 by Don Williamson?[26] Near Sydney, a flow of water usually referred to as the Tasman Front moves east, linking the south-moving East Australian Current off New South Wales with the southward flow of the East Auckland Current off Northland. Most breeding packhorse in Australia are found north of Sydney, and computer simulations using satellite flow data show that about 2 per cent of the larvae could be transported to New Zealand within a year.[27] However, the provisional results of a genetic study by Jenny Ovenden and colleagues point to the Australian populations of packhorse being different to those in New Zealand, in which case there can be no gene flow (although they stress that more genetic sampling is needed in order to be sure).[28] One explanation for packhorse larvae not being able to make it across the ditch may lie in the low biological productivity in the north Tasman Sea revealed by the low levels of chlorophyll there.[29] The larvae may simply starve.

To sum up, the way that packhorse recruit in New Zealand waters can be thought of in the following way (with information for closely related species incorporated to round out the account). The main breeding population is in the Far North. Hatching takes place there during December and January and the phyllosomas are dispersed well offshore. Most are carried generally east and southeast, those not being lost altogether from the New Zealand region having had their rate of transfer slowed by being caught up in eddies. The vagaries of the water flows, together with a measure of larval strategy, ensure that 9 months or so after hatching some larvae are to be found near the shelf break—the region at the outer edge of the continental shelf where the seafloor plunges from about 200 metres to depths of 1000 metres or more—at final stage. Final-stage larvae probably control their position in the water to some extent by directed swimming. They have large pleopods equipped with appendix internae (projections of the pleopod, which in pueruli bear hooks that link each pair of pleopods to strengthen swimming). Those with sufficient stored energy then metamorphose into pueruli.

The pueruli swim to the coast to settle in shallow waters, mainly along the east coast of the North Island. They don't disperse willy-nilly after metamorphosis, implying navigational ability. They probably use a combination of cues to

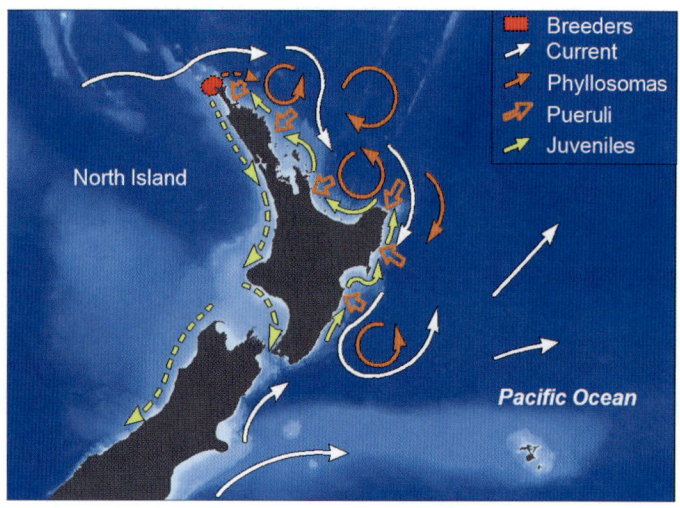

Recruitment mechanism for packhorse spiny lobsters in New Zealand. When sizing up a 5-kilogram packhorse in a crevice on the east coast of Northland, it is easy to overlook that this lobster was first, as a larva, carried hundreds or even thousands of kilometres in offshore currents, counter-currents and eddies over waters kilometres deep, then, as a juvenile, walked hundreds of kilometres to reach the spot below you.

orientate themselves, including low frequency sounds such as those produced by waves on the shore and the millions of kina (sea urchins) scratching and chewing on rocky reefs; reflected and refracted waves emanating from the shore; and the subtly different chemical and physical properties of the waters inshore versus those offshore.[30] The pueruli are active at night, so reducing their risk of being seen and eaten by visual-hunting predators. Their progress may also be assisted by mass shoreward water transport and wind-driven surface currents. The pueruli of many—if not all—spiny lobsters are capable of burying in sand.[31] This makes sense for, as they progress towards the coast after metamorphosis, they must take refuge on or very near the seafloor during the day to avoid predators. To bury, red rock lobster pueruli vigorously flick their tails to direct sand up into the water, leaving the tail partially or completely buried and a sprinkling of sand on the top of the carapace. The antennae are held spread apart, flat on the sand surface. (Interestingly, the very young juveniles can also bury like this, but neither they nor the pueruli tolerate fine silt.)

Settlement—which takes place in waters just a few metres deep—is a bit of a suck-it-and-see process which is not complete until the puerulus ceases any further forward swimming. Before that point the puerulus may have visited many potential spots in a process of active search and habitat choice at a small scale. After settling, and as they approach maturity, most of the juveniles then migrate north, back to the breeding area off Cape Reinga.

Not surprisingly, a similar mechanism of larval recruitment for *Sagmariasus verreauxi* is very likely in eastern Australia. Larvae hatched in the main breeding population, in northern New South Wales, are probably carried south in the flow of the East Australian Current, their progress slowed by being caught up in the embedded eddies that characterise this current. Settlement takes place in southern New South Wales (but at times even as far south as Tasmania) between September and January.[32] In New South Wales anyway, settlement is greater with distance south, perhaps because the shelf

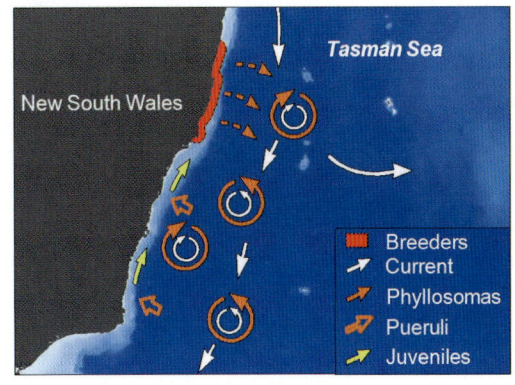

Recruitment mechanism for packhorse (locally known as eastern rock lobsters) in Australia. The distance between the main breeding area and the main settlement area is around 500 kilometres.

is narrower and so the larvae and postlarvae are closer to shore. As the lobsters approach maturity they migrate back north to the breeding grounds.

Other shallow-water spiny lobsters have an early life history very similar to that of our packhorse, characterised more than anything else by that bizarre-looking, transparent, leaf-like phyllosoma. But there is fascinating variety among species in precisely how the recruitment mechanisms work—as far as we understand them. Eddies, counter-currents, behaviour that changes as the larvae develop, migrations by juveniles—each makes an appearance according to species. The remarkable point they have in common is that all of these well-known, quintessentially coastal species must spend months and months well out of view of their home reefs. Development time varies according to species and sometimes area, from as few as 4 months (the ornate spiny lobster *Panulirus ornatus* of the tropics, and—surprisingly because it is a temperate species—the European spiny lobster *Palinurus elephas*) to between 12 and 24 months, possibly the longest for any invertebrate (the red rock lobster).[24] It is these long larval lives that give spiny lobsters their potential for wide dispersal and the opportunity to colonise new areas.

I can't depart this chapter on the strange and wondrous beginnings of spiny lobsters without mentioning a couple of intriguing New Zealand observations concerning the habits of that close relation of the packhorse, the red rock lobster. The first is an account by Noel Roydhouse published in 1988 in a journal that reports on, among others, diver mishaps. Entitled 'Diver's ear pain or claws two', it ended with the thought that spiny lobsters should perhaps be included among the dangerous marine animals.[38] His patient had for some time following a dive complained of a sore ear, which was later found to have within it a puerulus. Not surprisingly, the pain subsided after the lobster had been removed from the invaded orifice.

Then, in July 1990 at Castlepoint, on the southeast coast of the North Island, two of us in one hour collected almost 1000 pueruli and very small juveniles from an area of intertidal rocky shore a little less than half a hectare.[39] Many were alone in holes and indentations in the rocks, but up to 50 lobsters, and frequently more than 10, hid under individual rocks. Intertidal settlement of similar magnitude has been observed many times over the years at Castlepoint, but the reasons for this otherwise subtidal creature settling there so abundantly so high on the beach remains a mystery. Nowhere else does the red rock lobster settle intertidally in

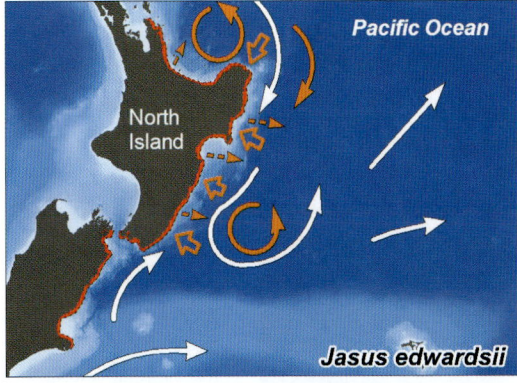
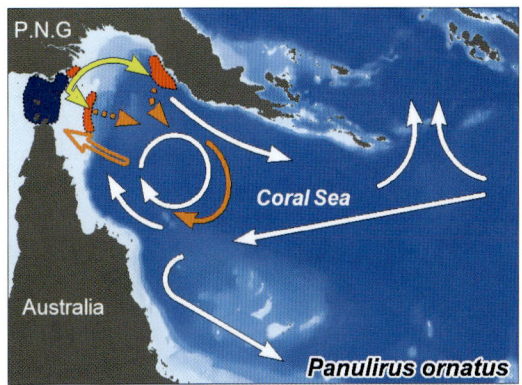
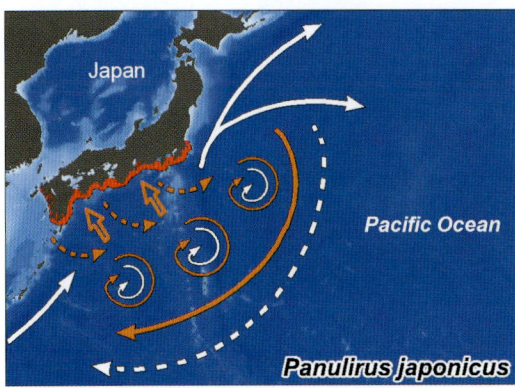
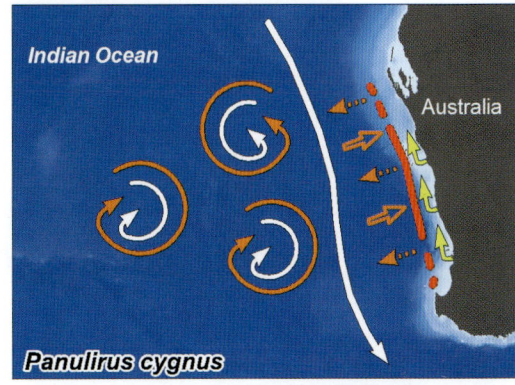
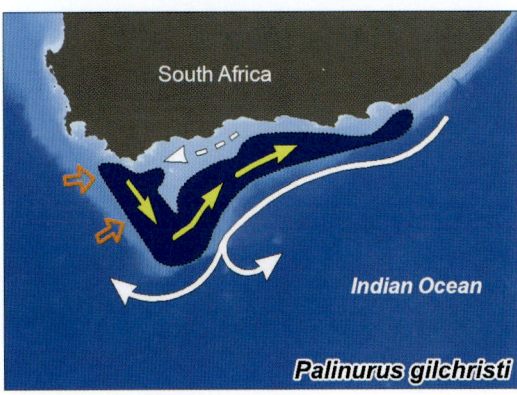
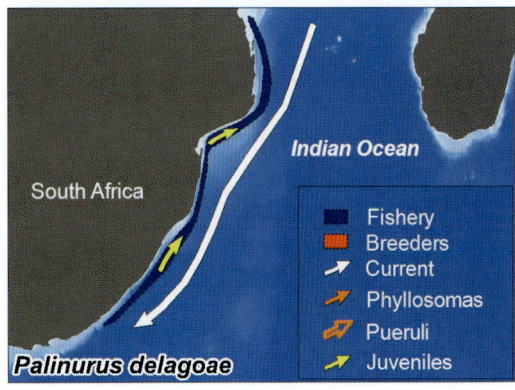

Examples of recruitment mechanisms among other spiny lobster species. Off the southeast coast of the North Island of New Zealand, red rock lobster *Jasus edwardsii* phyllosomas are restrained from being carried too far from parent grounds by being caught up in a huge eddy (upper left).[33] It is similar for the ornate spiny lobster *Panulirus ornatus* in the Coral Sea (upper right).[34] Phyllosomas of the Japanese spiny lobster *Panulirus japonicus* are thought to be returned toward hatching areas by counter currents (middle left).[35] Off Western Australia, western rock lobster *Panulirus cygnus* phyllosomas are returned towards parent grounds by spending most of their time at depths where the drift is shoreward (middle right).[36] Little is known about recruitment of the southern spiny lobster *Palinurus gilchristi* (lower left) and the Natal spiny lobster *Palinurus delagoae* (lower right) off southern Africa, but in both it is likely that juvenile migrations counter the drift of larvae.[37]

such numbers, and the counts are far greater than for any other species of spiny lobster. (The small lobsters make fine fodder for octopuses. There were 18 pairs of wee eyes in the stomach of an octopus found hiding one day in a shallow rock pool left behind as the tide had receded.)

There were close on 20 red rock lobster pueruli under this one boulder from the intertidal shore of Castlepoint in July 1990. Some are transparent, having only very recently settled; others show pigmentation of the new shell under the puerulus's one.

The beach where this highly unusual intertidal settlement takes place is just below the Castlepoint lighthouse on the open coast, directly above the gazer's eyes.

CHAPTER 5

A CLOSER LOOK AT PACKHORSE

As usual, their large size meant they were tricky to handle, but we had almost finished measuring a sample of packhorse caught a few days earlier near Cape Reinga. Then there was this true monster.

Smaller lobsters you can grasp with one hand to measure their bodies, like you hold a phone while texting. But it is very different with large packhorse. For the really big ones you need both hands—one each side—just to pick them up. And then you need something to support them while you measure.

It would not have happened had we been wearing those shiny-smooth plastic aprons people wear in processing sheds these days. Picking up this particularly large, rambunctious packhorse, Ministry of Agriculture and Fisheries scientist John Annala inadvertently allowed it too close to his chest—where the tip of a leg entered the weave of his woollen jersey. With an air of misplaced confidence he put the lobster on the nearby stainless steel bench, leaning over at the same time so as not to stretch the jersey too much. Using both hands he was now able to remove the offending limb. But by then the tips of two others had penetrated the garment, and so it went, one leg removed while two others engaged. Now the lobster was moving folds of captured garment towards its mandibles. Measured as its deliberate movements were, in no time this powerful packhorse was entirely embroiled in the front of John's jersey, in a kind of loving embrace.

There was only one solution: take the jersey off, lobster attached. After 10 or so exasperating minutes of finely coordinated extrication—me holding the lobster, John removing legs one by one, both of us striving to preserve won ground—we managed to separate the lobster from the jersey. But there was damage, not just ripped weave but holes chewed too.

At 10 kilograms, this lobster was large but not as big as they get. Packhorse are the giants among the spiny lobsters: males reach more than 280 millimetres carapace length (total length well over half a metre) and possibly 18 kilograms in weight, females over 260 millimetres carapace length and at least 15 kilograms.[1] Well up there among the larger bottom-dwelling invertebrates, their enormity and power must make them formidable to anything but the largest predator.

Nevertheless, packhorse represent all spiny lobsters—indeed all decapods—in their general appearance and body workings. They consist of two main parts. The cephalothorax is a fusion of the cephalon (head) and the thorax, the shell cover on the top being the carapace. At the front of the cephalothorax are the eyes and the antennae and antennules (feelers). Below them and back a little are the mouth and its associated feeding appendages. Beneath the cephalothorax are five pairs of pereopods (walking legs). Then there is the abdomen (tail), which consists of six segments, the last bearing the five-piece tail fan. Four of the tail segments have pairs of pleopods (swimmerets) beneath.

Many spiny lobsters are reddish in colour—red light is the first to be absorbed as light penetrates water. This is *Panulirus pascuensis*, one of the most remotely distributed spiny lobsters, known from only three places in the South Pacific Ocean.

The entire surface of the lobster is covered by the exoskeleton, mostly a thick shell. Composed of a chitin-protein material impregnated with calcium carbonate, it is structurally differentiated into a thin, non-chitinous epicuticle over a thicker, chitinous procuticle.[2] The different hues that characterise spiny lobster species are due to carotenoids (organic pigments), largely astaxanthin, in the pigmented layer just beneath the epicuticle. (There is, incidentally, a market for the shell. Both the chitin and astaxanthin are used in aquaculture feed and in other industries.)[3] To allow movement, the exoskeleton between the body segments and between parts of the

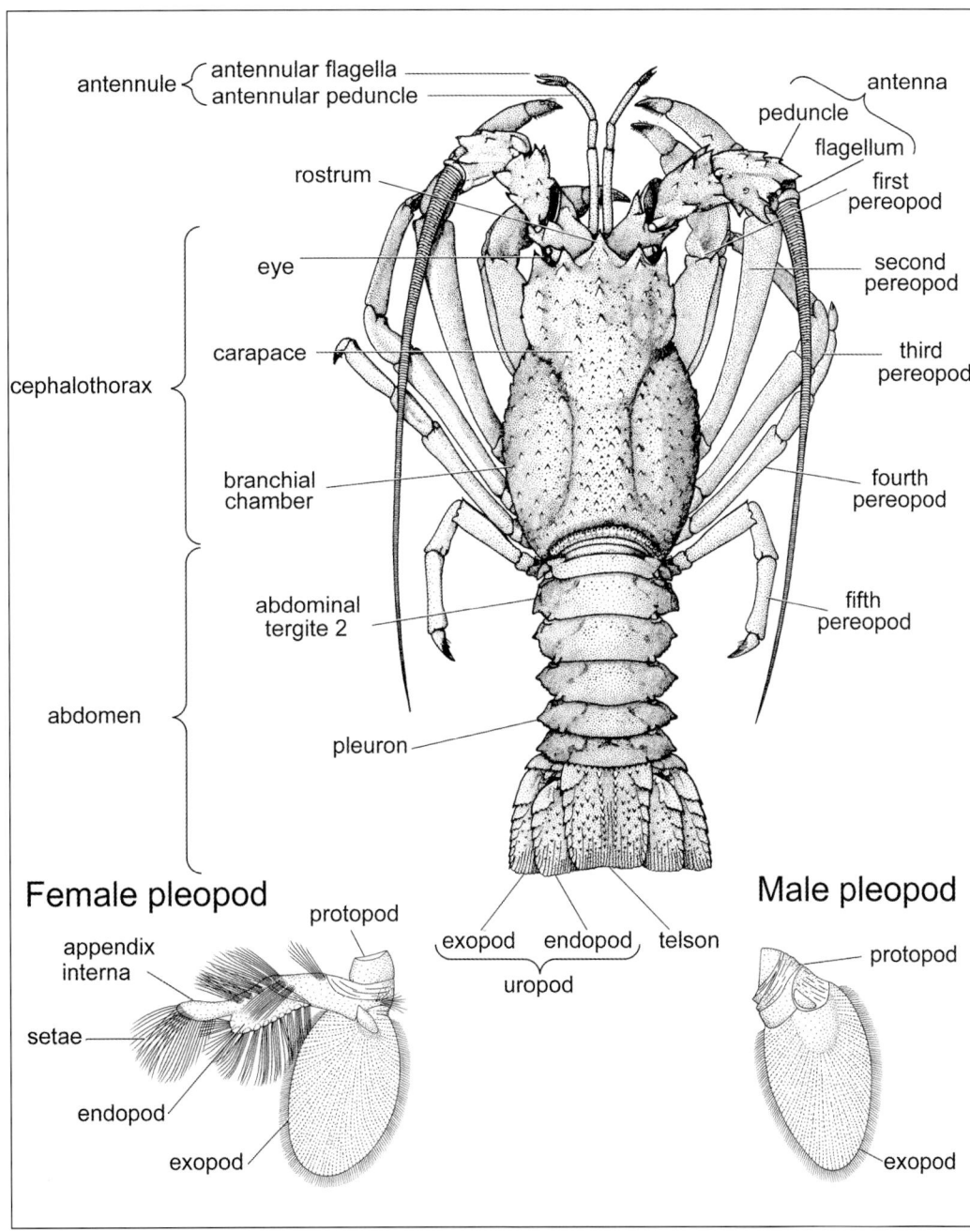

The external body parts of this male packhorse are typical of those of all spiny lobsters. Insets show a pleopod (swimmeret, located under the abdomen) from a mature female (left) and a male (right).

appendages is thin and flexible, these softer areas consisting mainly of epicuticle and uncalcified endocuticle.

Internally, packhorse share many features with other lobsters—and also with other animals including ourselves, even though their individual components may not be so evolved. They too have to deal with all those familiar things such as oxygenation of cells, digestion of food for energy and nutrition, elimination of waste, and reproduction. Much of the description of spiny-lobster body processes that follows applies particularly to packhorse but is based largely on Nellie Paterson's long-published—and still thoroughly relevant—detailed descriptions of a close cousin, the Cape rock lobster *Jasus lalandii*.[2]

The digestive system is of the filtering type, its tract being a more-or-less straight tube composed of three parts.[4] The first, the foregut, has a short oesophagus extending up from the mouth, and a large proventriculus that takes up most of the front part of the body cavity. The inner wall of the proventriculus has the gastric mill, with strong teeth which grind the food. It also has groups of setae (specialised hairs) which strain the food to ensure that only liquids and the finest particles pass on to the midgut. The midgut is a short, soft-walled chamber where absorption of nutrients takes place. Opening into it on each side is the duct from the large, yellow hepatopancreas (digestive gland). The last part of the digestive tract, the hindgut or intestine, is the longest. This straight narrow tube lies beneath the heart in the cephalothorax and extends the length of the abdomen to the slit-like anus on the underside of the last part of the tail.[2]

Once the food has been macerated by the mouth parts and forced into the mouth, it travels into the foregut where it is pulverised.[2] Here secretions from the digestive gland are added. The sifted material then passes into the midgut and enters the digestive gland, in the tubules of which nutrients are absorbed and stored. (The digestive gland also has other important roles in metabolism, detoxification and excretion.)[5] Larger undigested fragments are carried to the hindgut for elimination.

Ammonia is the principal nitrogenous waste product and is disposed of by several parts of the body: the gills, digestive gland, gut, body surfaces, and antennal glands. The paired antennal glands, which are at the base of the antennae on either side of the foregut, produce urine. Each has a bladder linked to an excretory pore at the base of the antenna. But, interestingly, the urine contains little nitrogenous waste, the main role of the antennal glands being ionic regulation of body fluids.[2] Also, much of the communicating that goes on between spiny lobsters is through chemicals released into the urine, as we will learn later.

Respiration is by means of 21 pairs of gills, each having developed as outgrowths of the limbs, as well as through various other surfaces of the branchial region. The gills are housed on each side of the cephalothorax in the branchial chambers, the 'packs' of the packhorse. Most gaseous exchange, resulting in oxygenation of the blood, takes place through the epidermis (surface layer) of the gill filaments. Water that aerates the gill filaments enters the branchial chamber through openings between the bases of the legs. Fringing setae screen out foreign particles. Passage of water around the gills is assisted by the movements of the gills themselves, but is mostly brought about by the scooping action of the scaphognathites (gill balers) of the second maxillae. The gill balers' strong sinuous strokes draw water from the branchial chamber, driving it forward and out through the exhalent aperture at the front of the lobster. At the same time, urine from the antennal glands is carried away.[2]

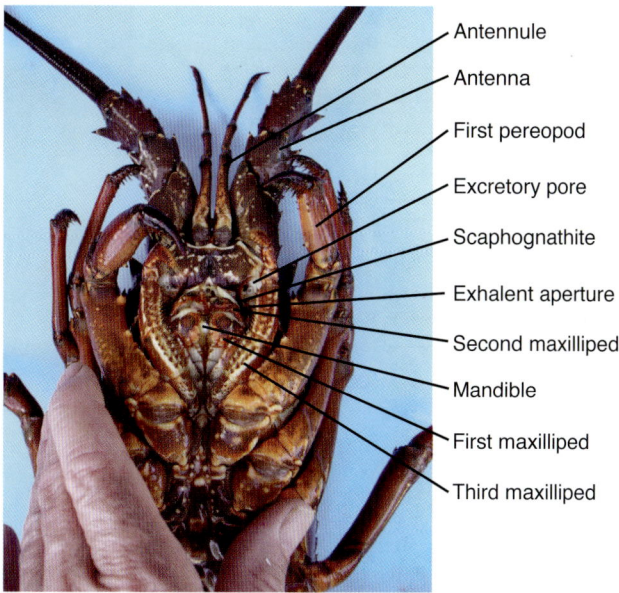

Underside of the anterior part of a female packhorse, showing the feeding appendages, mouth parts and excretory pore.

The circulatory system is 'open'—meaning that it is not completely separated from the general body cavity, or haemocoel.[2,6] Total blood volume is about 20 per cent of the body weight. Dissolved in the blood plasma is the copper-containing respiratory pigment haemocyanin, which has a bluish tint when oxygenated. The muscular heart—which beats once or twice each second—lies in the pericardial cavity, the space immediately beneath the central part of the carapace. Oxygenated blood is pumped from the heart into arteries which branch extensively and end in fine vessels in the tissues and organs. There is no direct capillary connection between the arterial and venous parts. Instead, deoxygenated blood collects in sinuses—irregular channels that are more or less continuous throughout the body and which are, in turn, connected to the main body cavity. Deoxygenated blood from all parts of the haemocoel eventually drains into the sternal sinus in the thorax, from where it is directed to the gills and gill filaments for oxygenation. After oxygenation, the blood returns to the heart through three pairs of openings and is once again ready to be pumped into the arteries.[2]

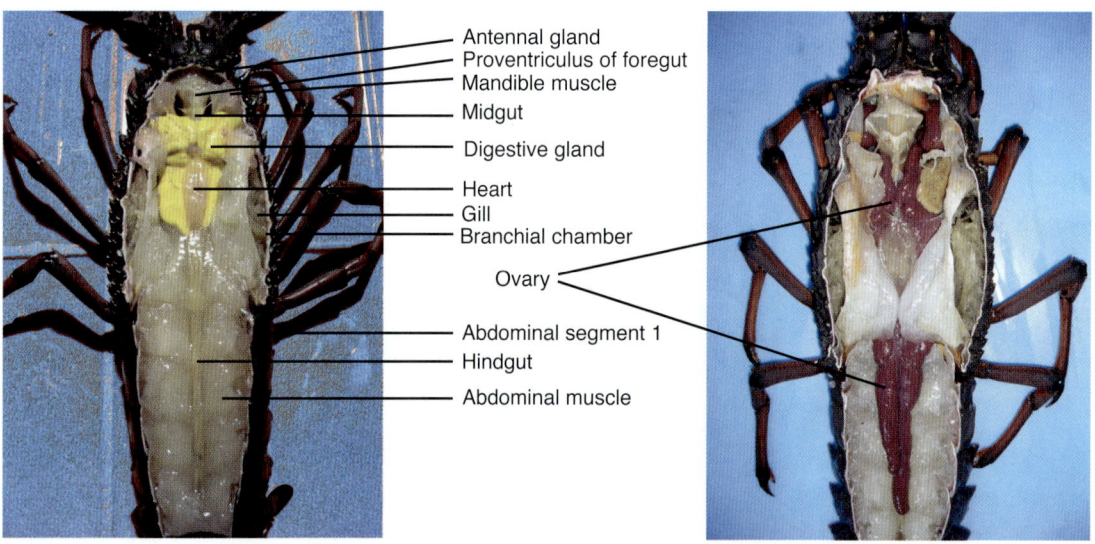

Views from above of an immature male (left) and a mature female (right) packhorse with the carapace, shell of the top of the tail, and the upper musculature removed. Both the immature female and the mature male closely resemble this immature male. In contrast, in a female about to extrude her eggs the red ovaries extend well back into the tail. Such is what you would find after performing this surgery on any spiny lobster.

The central nervous system is made up of a dorsal brain in the front part of the cephalothorax and a ventral nerve cord consisting of a chain of ganglia (small masses of nerve tissue) running the length of the body with nerves branching out to the muscles and other organs in each segment. Sensory setae (hairs and bristles) are widespread on the surface of the body and are either mechanoreceptors (touch-responders) or chemoreceptors (odour-detectors). The mechanoreceptors protrude as long, tapering feathered shafts all over the general body surface, but particularly on the antennal flagella and antennules. Chemoreceptors are usually shorter and non-feathery and are present on the antennae, the mouth parts, and particularly on the outer flagellum of the antennule.[2]

Mechanoreceptors are also found in the statocysts, which are paired balancing organs near the base of the antennules. These thin-walled, transparent, fluid-filled oval sacs are lined with fine mechanosensory hairs. Resting on the hairs are many statoliths (small sand grains) bound together by mucous. The hairs react to movements of the statoliths that result from the force of gravity.[2]

The eyes are stalked and compound, of the type typically found in arthropods that are night-active or which live in the deep sea, and they are particularly good at detecting motion. Those of at least the shallow-water spiny lobsters alternate between being light-adapted during the day and dark-adapted at night.[7] When dark-adapted, the screening pigments are so arranged that the eye glows when illuminated—just like cats' eyes at night. Viewed from above, the transparent cornea covering the surface of the eye is demarcated into facets, each representing an underlying ommatidium (ommatidia are the visual elements, each consisting of an outer corneal lens, a crystalline cone, and a rhabdome surrounded by retinular cells).[2] Light entering the eyes' many facets becomes focused on a single spot on the retina, greatly intensifying the image under low light conditions. It is unclear if spiny lobsters can discriminate colour but colour vision does appear to be widespread among decapods.[7]

The gonads are paired. The testes are found above the front part of the hindgut and in mature lobsters are long, white, highly convoluted tubes on each side, linked by a narrow bridge. The vas deferens runs from each testis to the aperture at the base of each of the last pair of legs. The ovaries are more conspicuous. They also lie above the front part of the hindgut but run further back, into the abdomen. The transverse bridge linking each side again makes them appear broadly H-shaped.

The oviducts are short, transparent tubes leading to the apertures at the base of the third pair of legs. As the eggs mature, the ovaries change in colour and shape, becoming bright coral-red and so engorged that they fill much of the body cavity, even reaching the bases of the antennae. After the eggs have been extruded, the spent ovaries are much smaller, collapsed, and greyish-white.[2]

After extrusion and fertilisation, the egg stalks twist intricately around the long setae on the endopods of the pleopods. The manner in which each and every egg—in the case of the packhorse, up to 2 million or so of them—becomes individually attached, firmly enough for them to remain fastened for several months, is both poorly understood and seemingly miraculous. The clue is thought to lie with the outer layer of the fertilised egg, which in seawater swells and becomes sticky.[8] Once glued to a seta, the egg is revolved by vigorous beating of the pleopods, so forming a twisted stalk which then toughens. With the egg anchored to and protected by the tail, embryonic development within proceeds until the naupliosoma (or, in the more evolved species, the phyllosoma) escapes by rupturing the egg capsule.

A remarkable characteristic of spiny lobsters is autotomy—the reflex severance of a limb by the lobster, equivalent to a lizard jettisoning its tail. Excessive bleeding is prevented by a valvular structure at the joint, the small quantity of blood that does exude forming a scab. Autotomy is useful to the lobster in avoiding being eaten and in limiting the effects of wounds. However, there are costs, both energetic (reduced growth at each moult and longer intermoult periods) and functional (less efficient foraging and mating success, and increased vulnerability to attack).[9] Autotomy is not particularly common among packhorse but it is a major issue in the western rock lobster *Panulirus cygnus* fishery off Western Australia. This spiny lobster often drops several—sometimes all—of its legs when handled, up to 80 tonnes of them being lost from the landed commercial catch each year.[10] This would be sufficient to dent anyone's wallet.

The packhorse fairly and squarely typifies the spiny lobster family in these broad aspects of body form and functioning. But it also displays characteristics not necessarily to be found in any other spiny lobsters. We can do little else than speculate on how these features—some unique in the family—came about. They result from earlier incarnations adapting to predation, competition and habitat change over the millennia. Nature has been referred to as a watchmaker working blind,[11] coming up with all manner of life-forms with an infinite variety of accoutrements, yet entirely incapable of learning from their adaptive success or failure. The virtual absence of a fossil trail, together with uncertainty about the ecological conditions that ancestral lobsters faced, mean that it is difficult to distinguish body features that were truly adaptive from those that were secondary, even random, developments.

The arrangement of the spines—together with the patterns of colour—are crucial in differentiating the various spiny lobster species. Spines give strength to the exoskeleton and help deter or thwart predators, and their arrangement also probably contributes to their owners blending in well with their backgrounds.[12] Colour is largely to do with camouflage too. Very young packhorse are first reddish brown, then dark green. As they mature they change to olive-green, becoming orange-yellow when old and large. These changes in colour are likely to be an adaptation to their changing habitats. The juveniles live in shallow, weedy places, but once they reach their breeding grounds in the north they are to be found on much more open grounds.

Both the fossil and living records of nature are scattered with examples of life forms increasing in size over time, bringing greater defensive prowess, access to a wider range of prey, and increased reproductive capacity.[13] Among spiny lobsters, the packhorse is the largest, both in absolute size and in the size at which it matures. The females also produce the most eggs in each batch, largely because of the gargantuan size they reach. Selection toward large body size, large size at maturity, and slightly smaller than average eggs, has led to greater egg production than any other spiny lobster. Packhorse in the Far North are probably about 8 years old when they mature after their long migration, and potentially they live to 30 years of age. It has been conservatively estimated that a female surviving to the maximum age produces close to 8 million eggs over her lifetime.[14] The large brood size and very high lifetime larval production suggest high mortality rates early in the lives of packhorse. In theory, there is balance between the levels of larval production and mortality; every female need produce only two lobsters (one male, one female) that reach breeding age in order to maintain a stable population. The enormous larval production of the packhorse possibly compensates for high mortality associated with their drift and settlement far to the south of where most of the breeders flourish, where conditions for growth and survival may be less than optimal. And the waifs still have to make it back to the adult grounds of the north.

The larvae of many subtropical and temperate-water invertebrates emerge during spring, which is when the plankton tends to bloom in the warming coastal waters. First to burgeon is the phytoplankton, followed by the grazing zooplankton and their predators; lobster larvae hatching at this time find themselves surrounded by food. In contrast, packhorse eggs are extruded in spring and hatch in summer. This is possibly because the waters of the Far North are usually at their most productive in summer, rather than spring, due to the seasonal pattern of upwelling.[15]

The larval life of 9 or so months for packhorse is round about the average for spiny lobsters—as far as is known. Existing for a long time as a larva provides opportunity for wide dispersal of a species, but at the same time the longer you drift around the greater your chance of being eaten or altogether losing your way. Some delicate balance—entirely mysterious to us—has been achieved by packhorse concerning the duration of their larval phase, the extent of their dispersal, and the numbers surviving the experience to eventually metamorphose, settle, and return to the breeding grounds. This balance is likely to be different to that of the red rock lobsters that live along the very same coast, for their larvae last twice as long in the plankton and yet are influenced by the same currents. The ledger of new recruits to the breeding population year by year is completely blank for packhorse, but we do have measures of recruitment for the reds, as we'll see in Chapter 8.

And, while on the subject of larvae, we may speculate on why packhorse phyllosomas have those extra appendages. Phyllosomas swim by rhythmically beating the exopods on their pereopods, and the third maxillipeds are used in feeding—immediately suggesting the advantages those additional limbs may bring packhorse. While swimming, the hairs on the exopods bend less with the backward stroke than with the forward one, bringing about forward movement. But because there isn't always an exopod on the third maxilliped, it's probably a characteristic on the way out rather than on its way in. Its presence is generally regarded as being a primitive feature.[16]

Packhorse have great propensity to migrate long distances—more than probably any other spiny lobster and possibly any other invertebrate that spends most of its life on the seabed. Such a strongly developed pattern of migration in

New Zealand is consistent with a local source of larvae rather than an Australian supply. Once having taken up residence on the seabed, the packhorse is at home on a wide variety of substrates—an essential adaptation for a species that must move over all manner of surfaces during its migration north and in the course of its seasonal inshore-offshore movements once in the Far North.

CHAPTER 6

DAILY ROUTINES

After their sojourn in the plankton so remarkably far from home, packhorse make this dramatic turnabout. They take up residence along the coastal fringe. Their transformation is permanent, from a weird constituent of the plankton into something ground-loving and much more familiar to us. There are still, nevertheless, many mysteries to their post-planktonic lives. Day after day, each lobster continues to establish associations amongst its own as it dominates, is dominated, migrates, feeds, mates—all the while trying to maintain a healthy detachment from those organisms that would devour it or invade it. Essentially none of this is witnessed, and little of it is well understood.

Naturally enough, our first enquiry concerns the youngsters, the new arrivals from the open sea. And, looking along its arc, this doesn't for a moment appear to be a good shell-collecting beach. There are too many rocks and banks of mobile pebbles, seemingly mutually exclusive to fragile life forms washed in from the sea. Facing directly east it has unhindered view of South America, with an equally unconstrained swell possible. Yet regular walks along this gravelly beach just south of the Bay of Islands, on the east coast of Northland, nearly always reveal one or two discrete lines of recently cast-up material. And along these bands, among both the natural (rosettes of bryozoans, eroded seaweed holdfasts, sponges) and the manmade (hopelessly tangled balls of thick nylon fishing line, plastic remnants of who knows what, an occasional bottle), can be found the exuviae (moults) of small juvenile packhorse. The severing low along one side of the carapace, and the broken membrane between the body and tail, indicate the route the lobster has taken to escape its old shell. Most exuviae are 4–8 centimetres in total length, although occasionally there is one so small it can't long have moulted from its puerulus. And it's all the more remarkable that they are to be found here at all because their brittle existence cannot normally be much more than a few days on such a harsh, exposed beach.

Few such Lilliputian specimens exist intact in New Zealand collections—Te Papa's assemblage amounts to just a small handful—and there has been no research directed into them in the wild. So, discovering that small exuviae can be found on this small beach—and not on others of very different character nearby—helps fill in a couple of lines of an otherwise very incomplete crossword concerning the habits of newly settled packhorse. The most important clue can be glimpsed on very low tides, when the packhorse's shallow, weedy, home reefs are plainly visible in the clear waters. Alison MacDiarmid of the National Institute of Water and Atmospheric Research (NIWA) once told me about disturbing a group of five small packhorse, each 3–5 centimetres in overall length, on similar shallow reefs at Mathesons Bay, on the east coast a little north of Auckland. They were associated with plants of the brown alga *Carpophyllum*, into the hearts of which they rapidly fled.[1] Dark-coloured, they seemed completely at home among the fronds—a far cry from the small red rock lobsters nearby, which were to be found individually in holes and crevices but never in the leafy parts of the plants the reef supported. Packhorse fisherman Adam Davey often sees little packhorse among the seaweed while diving shallow reefs near Long Beach, over the hill from Russell in the Bay of Islands.[2] Further, it seems almost inevitable that whenever one small packhorse is encountered, others are to be found nearby—although they may not necessarily be in close individual contact. Late one summer Daryl Sykes of New Zealand's Rock Lobster Industry Council stumbled upon an 'infestation' of them at a rocky headland at the south end of Waipiro Bay a little south of East Cape.[3] His shallow dive revealed dozens of them, 10–12 centimetres long, among the boulders and in weedy crevices.

When all the anecdotal information is considered in its entirety, you are led to the conclusion that small juvenile packhorse are most often to be found in shallow weedy waters, and that they seldom live alone. Seaweed usually means reef too, but it is the impression that the packhorse use the weed itself for cover that is distinctive. So it should not surprise us, then, to learn that the sampling equipment most effective in catching packhorse pueruli in New South Wales is made from artificial seaweed, and is moored at the surface.[4] Presumably the pueruli grasp seaweed—or its man-made equivalent—to steady themselves as they make their final move out from the plankton to become part of the seafloor community, and then for some time after they like to meld into the foliate structure.

The reddish-brown of the very young juvenile (left) turns towards dark green in older juvenile packhorse (right).

'Poorly known' best characterises our understanding of the ecology and behaviour of the very small juveniles of not only packhorse but virtually all spiny lobster species. The first few instars of the juvenile phase (or 'early-benthic phase juveniles' as they are more precisely known) are generally just too small, too sparse and cryptic, and the landscape of their habitat too complex, for them to be commonly encountered or routinely sampled. But generalisations can be made—at least for the shallow-water varieties. Whereas very small packhorse appear to be most at home in weedy situations, many other species prefer small holes in rocks or reefs into which their bodies will just fit. Most, if not all, have disruptive coloration, like army camouflage, and they remain in refuges, foraging only under the cover of darkness—and then only near their shelters. These precautions help them to avoid becoming fish tucker.[5] It appears that for most spiny lobster species, small individuals at this vulnerable stage of their lives don't care one way or the other about being close to others of their own. (Possibly packhorse are an exception in this respect.) Being asocial results in larger average distances between each little lobster, forcing predators to hunt for solitary individuals over a large area of habitat full of hidey-holes. The availability of such refuges can limit the local abundance of these juvenile lobsters, causing a demographic bottleneck.[5]

The point at which individuals become less clandestine and begin to seek out others of their own occurs abruptly at about 20 millimetres' carapace length, at least for the spiny lobsters so far investigated.[5] For algal-dwelling species, this means moving into creviced grounds where they can gather with others; for hole-dwelling early-benthic phase juveniles, the shift simply means vacating their small individual holes to occupy larger holes and crevices nearby. Den-sharing leads to more patchy distributions, corresponding to the distribution of larger crevices. It is also theoretically beneficial, because it enables group defence against predators. Not surprisingly then, this is when the juveniles begin to communicate using waterborne odours that attract others of their own, and it is also when the distinctive markings of the adult emerge.

Unlike ourselves, most spiny lobsters die young. Well-adapted as they are to their dark aqueous surrounds, the vast majority of phyllosomas perish through predation or starvation. And survival prospects remain grim for those lobsters reaching shore. It has been estimated that just 3 per cent of the Caribbean spiny lobster *Panulirus argus* in Florida survive their first year after settlement, with predators accounting for most of the deaths.[6] The pueruli of many species have been found in the gut of both midwater and seabed fishes, and they are also consumed by invertebrates such as octopus, and even by other crustacea such as crabs. Larger lobsters are also hunted by a wide variety of fish and invertebrates, including various cods, sea-breams, wrasses, scorpion fish, conger eels, small sharks, rays, octopuses, and other crustaceans—even penguins (apparently rockhoppers prey on the St. Paul rock lobster *Jasus paulensis*, although occasionally the tide has been turned and penguin meat used to bait lobster pots).[7] Growing rapidly to a large size is one way spiny lobsters help to remain unmolested and intact.[5] The very large (over, say, 5 kilograms), well-armoured adults probably have few non-human predators, with only large fish such as gropers and sharks, and mammals such as seals and sealions, being a threat. This is all the more so now that fishing has reduced the abundance of large fishes almost everywhere. Accordingly, large male spiny lobsters, particularly during the breeding season when they tend to cast caution to the wind while trying to attract a mate, will quite happily move about over open terrain on their own, even during the day.[8] Of course, they remain tightly holed-up during and immediately after the moult, when they are soft-shelled and vulnerable.

The list of tactics used by the predators of spiny lobsters is extensive. Some are night workers, others operate only by day. A number hunt and chase, others ambush, and then there are those that probe and grab. The cues to the whereabouts of the lobster are not only visual, but also hydrodynamic and chemical.[5] To foil these tactics, spiny lobsters are camouflaged and they hide, they escape by backward swimming, and they may defend their position, the effectiveness and importance of each strategy changing during the lobster's development.[8] Postlarval and newly settled lobsters are vulnerable to the widest range of predators, and remaining hidden as much as they can is their trump card. As they move away from their settlement areas, escape—especially by repeated tail-flipping which propels them rapidly backwards over many metres—together with crypticity, are important. As lobsters grow larger, their suite of potential predators to be afraid of decreases dramatically, particularly because their armour is now so much better developed and more effective. This is also when the benefits of den sharing and joint confrontation become a driving force for aggregation.[8] Crevice-sharing reduces the chance of detection. It also helps protect the incumbents from attack once discovered: when bunched tightly together, fending off predators by whipping and thrusting their antennae, spiny lobsters can be formidable, as any diver will attest.[5] At the same time, lobsters forage alone at night to reduce the risk from visual predators, and they may use chemical cues to detect and avoid predators such as octopus.

And spiny lobsters are themselves predators, becoming 'restless' as sunset approaches, and emerging from their shelters shortly after. They typically forage up to a few hundred metres from their 'home' shelters before returning near dawn.[8] They dine widely and opportunistically, but also choosily where there is choice. Although most food in the gut is pulverised virtually beyond recognition, it is now known from forensic analyses of the pulp that many if not most spiny lobster species consume molluscs, crustaceans, echinoderms and other invertebrates, and also various amounts of coralline and fleshy algae according to species and place.[5] Given the opportunity, though, high-energy items such as shellfish are preferred over others. And the prowess and determination of spiny lobsters to detect and then deal with such prey items as bivalves are remarkable. Like slipper lobsters—whose feeding practices have been closely examined—they probably use chemoreceptors on their antennules to detect and orientate toward the chemical cue emitted by the prey. With the mollusc removed from the substrate, the lobster must then somehow separate its pair of shells. Typically it uses its mandibles to crack and chip away at the shell margins. Once a sufficiently large hole has been gained, the lobster either digs out the flesh with its leg tips, or it applies sufficient brute force over time to break the hold of the shellfish's adductor muscles.[9]

Large, long-lived, mobile predatory invertebrates such as spiny lobsters impact profoundly the ecological character of the grounds on which they live. They are 'keystone predators'—predators whose impact on the ecosystem is disproportionately large relative to their abundance. And because they feed high in the food chain, lobsters can initiate changes in the shape of the communities in which they live. These 'trophic cascades' occur when such predators suppress the abundance of their prey, thereby releasing the next trophic level from predation (or herbivory if the next trophic level is a plant-eater). For example, spiny lobsters, together with predatory fish, often control the abundance and distribution of sea urchins. The grazing activities of these urchins, in turn, affect the abundance of kelp and hence the character of kelp forests. Take away the lobsters and you end up with a burst in urchin abundance—which can hit the kelp hard. Such is seen in northeast New Zealand where top-down control by predatory fish (mainly snapper) and the red rock lobster on kina populations in turn influences the abundance and distribution of habitat-forming seaweeds.[10] When numbers of snapper or spiny lobster are reduced through fishing, 'urchin barrens' result, as out-of-control urchins devour practically the entire seaweed cover.

Kina (sea urchin) barrens. When predators such as spiny lobsters are removed (usually by fishing), the sea urchins begin to exhaust the seaweeds on which they graze.

Levels of cannibalism are probably low for most spiny lobster species in most areas, but where food is short it can bring about appreciable levels of mortality. In a recently established marine reserve on the east coast of the North Island of New Zealand in which there was a dramatic increase in the numbers of red rock lobsters, Department of Conservation's Debbie Freeman saw not only changed community structure but also significant levels of cannibalism, as discussed later.

The oft-asked question concerning a large spiny lobster, 'How old is it?' may never be definitively answered. All hard parts are shed at the moult, so the lobster has no equivalent to a fish's ear bone to section as you would a tree trunk for evidence of annual rings. Even the lining of the digestive tract goes. In the absence of conspicuous age marks we must look to established age-to-body-size relationships and the factors that contribute to its variability, together with computer modelling of growth. Using such data, it has been estimated that packhorse live for 30 years—given the chance.[11]

Age in spiny lobsters is usually taken from the time of settlement, the duration of the larval phase being not well-enough known or too variable to be included. And, like all spiny lobsters, packhorse start the first few days of their life on the seafloor pretty small. The puerulus is about 11 millimetres in carapace length (22 millimetres total length excluding the antennae) and weighs about 0.5 grams. Even after the first moult it is still only 16 millimetres in carapace length. But spiny lobsters in general—and packhorse in particular—are among the fastest-growing and largest of all crustaceans.[5] Most species attain 1–10 kilograms in weight and 30–40 centimetres in overall length; packhorse, of course, achieve substantially more. And because survival, reproduction, migrations and so on are age-dependent, and in turn related to body size, knowledge of growth is vital for coming to terms with the population dynamics of lobsters, and to their sustainable management. Maturation brings with it slower growth, especially in females because they must

This small red cod had consumed a puerulus and young juvenile red rock lobster.

allocate more energy to reproduction than males do. Species that mature early and continue to grow rapidly are more productive than those that grow slowly and mature late. And packhorse? They mature rather late in the piece: although they initially grow quite quickly, they don't mature until they are very large (about 160 millimetres carapace length) and quite old (about 8 years).

Because the old shell must be shed in order for the lobster to grow, growth is not smooth and continuous but is, instead, a stepped process—often with a great deal of variability in step size between individuals. (Moulting takes place in a sequence of stages in which the old exoskeleton separates from the underlying epidermal cells and new cuticle is formed. After the old exoskeleton is shed, the cuticle thickens and hardens.)[12] Growth rate has, therefore, two distinct components: moult increment (how much the lobster changes in size each moult, which—particularly in large individuals—can be negative) and moult frequency (how often the lobster moults over a particular period). It is the course of early growth that is usually best known—allowing reasonable estimates of the time taken to reach maturity, legal size and so on. Far fewer large, old specimens have been tracked for their increment and frequency of moult. This is because most tagging projects rely on commercial fishers and it is expensive to first purchase a large lobster so it can be tagged, and then to repurchase it after recapture.

So how does the growth rate of packhorse in the wild compare with that of other spiny lobsters? Based mainly on New South Wales field data, packhorse 3 years after settlement have a carapace length of 80 millimetres; at 6 years they are 140 millimetres, and at 10 years 190 millimetres.[11] This rate of growth is faster than that of red rock lobsters in New Zealand and the European spiny lobster *Palinurus elephas* in the Mediterranean (60, 90 and 120 millimetres carapace length respectively). But it is well short of that achieved by tropical species, such as the Caribbean spiny lobster in Cuba (120, 170 and 200 millimetres carapace length).[12] Always, though, there tends to be a lot of variability in growth rate between the sexes, between regions, and even between years, to say nothing of individual variation.

Packhorse reproduce once a year, during spring and summer. The females are unusual among spiny lobsters in that there is a delay of about 2 years between the appearance of setae on their pleopods (usually a sign of maturity) and the first time they bear eggs on those setae. This phenomenon was first revealed in the catch samples that the then Marine Department's Craig Kensler collected in Spirits Bay in 1966. His maturity-frequency graph for October, before the females had egged up, looked right. There were two categories of female: the immature ones up to 170 millimetres carapace length; and then those that were larger, with full-length setae on their pleopods to which the eggs attach—and which were presumably mature. But in December, there was a third, anomalous group of females. Between the smaller immature females and the large berried ones was a sizeable contingent of individuals that had well-developed setae but were without external eggs.[13]

A single catch sample—curious, but perhaps nothing to be made too much of. But other catch samples taken during other summers, and at other places quite well distant, showed very similar maturity distributions to the December 1966 Spirits Bay one. Was there for packhorse a size difference—and therefore a time delay—between first appearance of setae and the first time they carried eggs? Or was there some other explanation, such as the smaller setose females being elsewhere at mating time and so missing out on being fertilised; were they unavailable for mating because of the stage of their moult cycle; or were only large females in the area being mated?

Although having fully developed setae on their pleopods, the smaller female packhorse in New Zealand are still either functionally or physiologically immature.[14] Most females are at least 160 millimetres in carapace length, and about 8 years old, before they first bear eggs. In only one other species—the western rock lobster *Panulirus cygnus* of Western Australia—is there known to be such a delay. The delayed *P. cygnus* females are referred to there as 'inactive breeders', and their ovaries have been found to be undeveloped, without mature egg cells.[15]

Size at onset of breeding varies a lot between spiny lobster species.[14] Generally, warm-water species are younger and smaller at maturity than cool-water ones. The spotted spiny lobster *Panulirus guttatus*, a tropical species that seldom reaches 80 millimetres carapace length, probably has, at 40 millimetres carapace length, the smallest size at first breeding of any female spiny lobster. In contrast, the size at onset of breeding in the cool-water *Jasus* species can—depending on location—be more than 110 millimetres carapace length. Bucking this trend, at 160 millimetres carapace length, the subtropical packhorse has by far the largest size at first breeding of any female spiny lobster. (Maturity in male spiny lobsters is much more difficult to pin down than it is for females, usually requiring dissection. But where it has been investigated, males have tended to mature at a similar size to the females in any particular area.)[14]

The geographical variation in female size at onset of breeding in many spiny lobster species is often perplexing.[14] For example, red rock lobsters on the east coast of New Zealand from Banks Peninsula south (and including Stewart and Snares islands) have a much larger size at maturity than elsewhere.[16] There are no known genetic differences to account for this variation, nor is there slower growth in the areas where first-time breeders are small. The differences have generally been attributed to the cooler water in the region, but this explanation is not entirely satisfactory: waters off Kaikoura are just as cool as they are further south, yet size at first breeding there is small. But not all instances of variable size at onset of breeding are as mysterious. Density-dependence—where the biological response depends on the density of lobsters present in the area—can, by affecting growth rates, strongly influence size at first breeding. Negative growth (shrinkage) in the Cape rock lobster *Jasus lalandii*, probably caused by reduced food availability resulting from a large-scale environmental perturbation associated with the El Niño years of 1990–93, significantly reduced the size at onset of breeding.[17]

Spiny lobsters must moult before they conjugate so that the setae are long and silky—not short and ragged—for the attachment of the new batch of eggs. Clawed lobster females are usually still soft from the moult at mating; spiny lobsters have hardened up by mating time, but still there can't be too long between moulting and egg laying or there will be insufficient time for the eggs to develop before the next moult is due.

Spiny lobsters have a surprisingly complex mating system, our understanding of which has developed a lot through the work of NIWA's Alison MacDiarmid. Her following account of mating in the red rock lobster provides a template for packhorse and other spiny lobsters,[18] even though details vary to some extent between species, particularly with regard to the longevity of the sperm packet.

> Both sexes play a role in deciding who mates with whom. Large males become aggressive during the mating period and push other males out of their shelters, usually leaving only one large male per den. In extreme cases where males are a similar size, they rear up and fight one another, each using his powerful front pair of legs to grip and crush. These embraces may last

for 1–2 minutes until one male releases his grip and shrinks away, leaving the victor standing tall. Where large, 2–3 kg males are still common, this aggression excludes smaller males from participating in reproduction.

Large males prefer to mate with large females. Large females have many more and larger eggs than small females, and the eggs hatch into larger larvae that are more resistant to starvation. While males are choosing females, the females themselves are busy choosing males to shelter and mate with. In the days or weeks between moulting and mating, females shelter with the largest males available. Small males cannot usually provide large females with enough sperm to fertilise all their eggs, so there is a strong incentive for females to select large mates. If there are no males available to mate with, females usually reabsorb the eggs in their ovaries; the resulting ovarian scars lead to reduced clutch sizes the following year.

Once a mate has been selected, the [lobsters] court for several minutes or for as long as 2–3 days before mating. The male faces the female, close enough to enable both [lobsters] to touch antennules (a) and b) in the illustration below). The lobsters chemically signal in their urine their readiness to mate and this is wafted toward the partner. They rear up belly to belly and embrace tightly (c). The pair normally overbalances, landing with the female uppermost and the male's rear walking legs now tightly embracing the female (d). Within 90 seconds the male deposits an external packet of sperm (from pores at the bases of his last pair of walking legs) onto the belly of the female (e). The sperm packet starts to disintegrate immediately so the female rapidly extrudes her eggs. She usually clings in a vertical, head-up position to a nearby rock face and forms a brood chamber by flexing her abdomen forward beneath her body, extending the pleopods and spreading her tail fan to cover the genital apertures (at the base of her third walking legs) and the sperm mass. The extruded eggs are fertilised as they pass down into the brood chamber and are attached to the long hairs on the pleopods. Egg laying and attachment usually takes less than 50 minutes.

Fertilisation in spiny lobsters is therefore external. The sperm transfer organ is the gonopore itself, the sperm being transferred en masse to the female sternal region. In both packhorse and the *Jasus* species, the sperm packet, or spermatophore, remains intact for just a matter of minutes, so the female must lay her eggs immediately. But in the more evolved genera, such as *Panulirus*, the spermatophore takes the form of a putty-like mass ('bog' or 'tar spot') that can exist for weeks.[14] When ready, the female breaks the surface of the bog with a

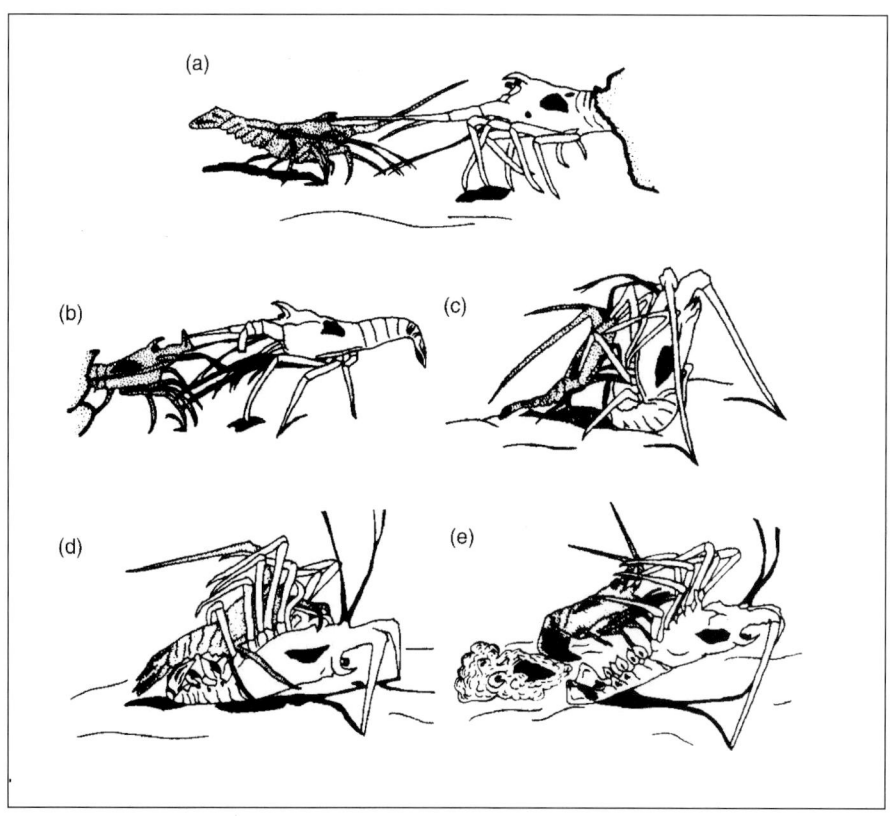

Mating spiny lobsters, the female being the darker one.[8] Labels refer to the quotation on the previous page.

leg-tip before extruding her eggs. It is probably general that smaller males pass less ejaculate than large males, and clutches fertilised by small males are smaller than those fertilised by large males. Large males may deposit a spermatophore of a size proportional to that of the female. Females normally mate only once before each successive brood but more profligate ones, particularly those that have been mated by a small male, will occasionally remove a spermatophore.[14]

External fertilisation is usually thought of as being inefficient compared with internal fertilisation because much of the sperm is lost to the watery surrounds. And shortage of sperm—too few males, or males too small to provide sufficient sperm to fertilise all the eggs of all the large females present—can become an issue in spiny lobster stocks that are fished. The problem is compounded by sperm-to-egg ratios often being relatively low and sperm regeneration slow.[14] Sperm limitation is likely to be first noticed in heavily exploited fisheries, particularly those where the males are more heavily fished than the females—and so are generally smaller—because they do not share the protection from fishing that egg-bearing females enjoy.[14]

In most spiny lobsters, body size is the main determinant of brood size. Typically there are, according to species, between 100 000 and one million eggs, each 0.6–0.8 millimetres in diameter, in every brood.[14] Packhorse are so large that they produce far more eggs per brood—400 000 in the smallest, and up to two million or so in the largest. Temperate-water species—as well as subtropical ones such as packhorse—have one brood of eggs each year and a relatively long brood period of several months. In contrast, tropical species generally have brood periods of merely a month or so, and often they reproduce more or less continuously throughout the year. Nevertheless, the average potential life-time egg production of packhorse, conservatively estimated at 8 million, is almost certainly the greatest of all spiny lobsters.[19] This gargantuan egg production is, as discussed in the previous chapter, likely a response to high mortality associated with wide larval dispersal and late breeding, as well as the extensive migration of juveniles that takes place before they reach their breeding grounds. In contrast, among those with the most modest life-time egg production is the diminutive and small-maturing spotted spiny lobster, at 0.3 million.[19]

The female keeps her abdomen flexed forward for most of the incubation period, creating a protective enclosure for her eggs. The pleopods beat slowly

Intact spermatophore, or tar spot (the black object on the rear part of the sternum), on a female western rock lobster. When ready to extrude her eggs, she will rupture the spermatophore to allow her eggs to be fertilised before they attach to the setae on the pleopods beneath her tail.

to oxygenate the eggs, and periodically she extends her abdomen to give her eggs a once-over and to preen them with the small pincers on her last pair of walking legs. This is when unfertilised or diseased eggs are removed. When her eggs are ready to hatch she will, at sunrise, repeatedly stand on the tips of her legs and extend her abdomen vertically into the water current, violently beating her pleopods for a few seconds. This releases a 'swarm' of naupliosoma larvae that swim strongly towards the sunlit surface waters.

Packhorse in the Far North at egg-hatching time are to be found in tide-swept offshore waters, so it is not hard to imagine their larvae being quickly dispersed to the high seas. But what about more inshore, crevice-dwelling species? Larvae hatching from lobsters in nearshore fissures would be vulnerable to reef predators and difficult to disperse without access to more open-water currents. NIWA's John McKoy and Andrew Leachman reported large, egg-bearing red rock lobsters in areas of strong tidal water movement during late winter. In the absence of suitable shelter, the lobsters took the form of defensive circles or heaps.[20] The animals in the circles faced outwards, their antennae angled back at 45°. Being out in the open like this is hazardous for a crevice-dweller, which suggests that the risk of her newly hatched larvae not being effectively dispersed is greater than the risk of the mother becoming fish fodder.

Gregariousness and keen sociality characterise the adult lives of most, if not all, spiny lobsters. These qualities are most obviously to be seen in the sharing of dens, which maximises survival through group vigilance and cooperative

A red rock lobster with hatching naupliosomas.

Large male red rock lobsters such as this one are valuable to fishers, but also to the stocks—because large females require a large male.

defence.[8] But the use and re-use of particular dens depends on how much of a rover the species of lobster is. Given adult packhorses' migratory propensity—they are more or less constantly on the move, marching from waters only metres deep out to those closer to 200 metres and back again, according to season—it is unlikely that this species occupies specific dens for long spells. For the less itinerant species, there is a tendency to occupy only some of the available dens within an area—even though others appear equally suitable—and to occupy them for weeks or months, even returning to the same dens after periods away. It is possible that the lobsters mark these dens as 'safe houses', using long-lived chemicals in their urine.[8] Also, once one lobster has arrived at a suitable den, more are likely to follow. Around 40 per cent of red rock lobsters in northern New Zealand have been found to occupy the same primary dens over long periods.[21] Some lobsters in the Cape Rodney–Okakari Point Marine Reserve have been resighted in the same place 8 years after having been tagged there, although of course they will have visited other dens too.[22]

Although spiny lobsters see well—their eyes are efficient at detecting low light levels and perceiving motion, and so are essential in social interactions as well as in predator avoidance—it is odour that is most pervasive in their social dealings. The chemical signals the lobsters both emit and detect are able to bring about very disparate behaviours.

Marine animals generally don't need a urine bladder as they can release wastes directly into their environment as quickly as they are produced, so for a long time scientists were puzzled by the fact that decapods stored their urine. We now know that spiny lobsters release pheromones (chemical signals) with their waste. Communicating by means of your urine may seem distasteful to us, but it is a perfect strategy for the lobsters. The urine is released through a pair of nephropores, small holes at the bases of the antennae. Gill currents, produced by the large gill bailers pulsating within the gill cavity as the lobster breathes, carry the urine forward into the water column, to be detected by chemoreceptors on the antennules of any nearby lobsters.[23]

These chemical messages influence all manner of behaviours—amorous, combative and conciliatory. Natalie Raethke at NIWA found that in the red rock lobster, for example, these pheromones are particularly important in mate selection.[23] During the short reproductive period, from the end of April to the beginning of June, large post-moult female red rock lobsters were highly attracted to *large* males, not small ones—mediated by their different odours. She also found that only during the mating season did male rock lobsters fight for dominance, often not tolerating any other male close by. In extreme cases, a fight would result in the loss of limbs or even the death of one male, so the loser of a fight usually avoided further confrontations with the same dominant male. It appears that the male lobsters not only recognise each other for a time with the help of pheromones, but that they can also signal to a female whether they are winners or losers. The female then usually chose a dominant animal to mate with.

Within species of the Stridentes, sound may also be used as a form of communication, possibly with a reproductive function.[8] However, the primary role of the stridulating (sound-producing) organ seems to be to startle predators. The Caribbean spiny lobster emits sound with rapid whips of its antennae when attacked by triggerfish.[8] Further, those that have had their sounds experimentally muted are less successful in avoiding predation by octopuses than those capable of noise.[24]

Lobsters that aren't killed by predators must, nevertheless, avoid afflictions if they are to live a full lifespan. But in the wild they appear to be a pretty healthy lot. The most widespread problem among spiny lobsters is erosion of the shell and associated tissue—bacterial shell disease.[25] Usually non-fatal, it is most common in lobsters that have been stressed. The infection begins as small 'burn' spots in the cuticle which then merge to form continuous eroded lesions. The condition often arises from scuffing of the exoskeleton, so, not surprisingly, multiple handling in intensive fisheries can increase its frequency. The sites of lesions vary, but in several spiny lobsters, including packhorse, they are most often the uropods and telson—with the condition often being referred to as 'tail fan necrosis'. Usually individuals overcome bacterial shell disease by moulting, so larger lobsters which moult less frequently—particularly mature females—tend to be those most affected by it. Their unattractive appearance, together with their increased morbidity, reduce their market value.

Turgid lobster syndrome is a poorly understood affliction in which membranes of the tail segments become inflated with clear, watery haemolymph.[25] The lobster becomes lethargic and stops feeding, and, in up to 50 per cent of cases, will die. Most commonly seen in captive lobsters, it is also found from time to time in wild populations. Salinity change or suboptimal water quality may contribute to its outbreak.

Whitening of lobster flesh is usually caused by a microsporidian parasite that invades the muscle tissue and forms large numbers of microscopic spores. It can be fatal. Other infestations—including various crustaceans, fungi, protozoans and flat worms—are seldom any real problem for spiny lobsters in the wild.[25]

But nature is never entirely predictable. The recent appearance of the PaV1 virus in the Florida populations of the Caribbean spiny lobster was a shock; it was the first report of a viral disease in spiny lobsters in the wild.[5] One of the few known to have a high incidence in nature, this virus is nearly always lethal, although lobster susceptibility appears to decline with age. Interestingly, because healthy individuals can identify those diseased and will not join them, the virus alters the spatial distribution of juveniles.

Bacterial shell disease in packhorse and other spiny lobsters is often seen as 'burns' on the shell (left) and erosion on the tail fan (middle). Occasionally the tail erosion spreads so much that it becomes unsightly (right).

CHAPTER 7

THE PAGEANT OF FISHING

Fishing has irrevocably changed the character of all but the deepest and most remote of spiny lobster populations. Our fondness for them as food ensures that we are inseparably part of their ecology. But this is not completely without reward because the very act of fishing them has revealed some of their interesting habits.

Adam and Nat Davey commercially fish packhorse in the Far North on their vessel *Medea*—with considerable success. In early 2007, I watched their catch from a single lift of 50-odd pots near Spirits Bay being landed 150 kilometres to the south at the tiny Bay of Islands port of Russell—known during the whaling days a century and a half earlier as 'the Hell Hole of the Pacific' for its bars and debauchery.

Medea is wooden, close on 20 metres long, sheer-lined with an expansive afterdeck. Adam and Nat began to transfer the lobsters from enormous blue plastic bins, each holding a tonne or so of seawater, to smaller plastic fish boxes that could be heaved to the wharf above. What particularly excited the tourists on the wharf was the sheer size of these lobsters, most of them packhorse and some close to half a metre long and several kilograms in weight. The larger of them had that characteristically well-developed pack-like branchial region. The crew handled the catch charily, not just because the fewer missing limbs the lobsters had the more they were worth, but also because they were not wearing gloves and the lobsters' flailing feelers can inflict painful abrasions. There was a clattering rumpus from the plastic bins as the lobsters flicked their tails in attempts to escape. The more subtle, haunting sound was them grinding their mouthparts.

Packhorse fishing vessel *Medea*'s home port is Russell in the Bay of Islands.

The contrast in colours was striking: the smaller lobsters were a strong dark green, like silverbeet in full season; the larger ones were more olive, some almost light brown.

'Not a bad catch. It's hard to see how this much could come from a fishery that is supposed to be poked,' mused the skipper's father, Ross. This fishery had, when considering the landing figures on their own, performed rather poorly over the previous couple of decades. Yet here it was yielding something like a tonne of lobsters from one lift of just a few dozen pots.

These lobsters were bound for Auckland markets, particularly to be centre stage at Chinese New Year celebrations for an ever-expanding, wealthy and discerning Asian clientele. Each of the larger ones would cost the diners well over a hundred dollars. But as impressive as this landing was, it was a far cry from the truly huge individual landings—each of many tonnes—of large packhorse of the 1960s and 1970s. Most would go to Happy Yovich at Hikurangi Fisheries in Hikurangi, a few dozen kilometres inland and south of Russell. There the packhorse were processed, the tails being frozen and exported, mainly to the United States. A proportion of the many huge bodies were cooked and sold chilled at the factory door or taken down to Auckland to an eager, mainly Polynesian market. The rest was dumped.

φφφ

On State Highway 1 a little north of Kaitaia a lone backpacker stands on the roadside, hitching to Cape Reinga. 'See if she wants to come by sea,' Trevor suggests. From Scotland, she makes it five for the February 1976 trip to the extreme northern tip of mainland New Zealand where *Mako*'s skipper Trevor Hare pots packhorse. The 23°C sea-surface temperature reveals the swathe of subtropical water pressing down on Northland. Flying fish are disturbed by our wake, escaping their assumed pursuer in a fluttering shimmer.

While sheltering overnight in Spirits Bay ahead of the first lift of the pots, the backpacker enquires 'Do you often get to eat the crays yourself?'

'You always look forward to the first trip of the season,' replies Trevor, sipping the ubiquitous sweet, strong tea offered in the wheelhouse of any small fishing vessel. 'There are always a few undersized or damaged ones. You eat the lot, the mustard, the tail—preferably seared in garlic butter. The next night, after you've anchored and settled for the night, it's just the tail you're yearning for—maybe mornay. Later in the week, it's just the head meat—particularly the meat at the base of the feelers. Next time? You try the bait.'

Soon after dawn *Mako* is well offshore, drifting, and there is not a buoy to be seen. Te Paki, darkly backing Cape Reinga and at 310 metres the highest point on the often precipitous northern tip of mainland New Zealand, is a mere shadow on the horizon to the southeast. The Three Kings Islands are scarcely more visible to the northeast. The sea is dark and oily and strangely forbidding, the angle of the early sun too oblique to penetrate its transparent mass. Suddenly red floats begin to pop to the surface all around. Some disappear just as quickly, the tidal currents taking them down again in a powerful lick. It's close to slack water, when the tide stops running before reversing its flow, and so eventually the floats will win. But it will be only 90 minutes before once again they are pushed to the depths, irretrievable for the following 5 hours. So things happen fast.

Trevor steams to the first float and the crew casts a small stainless steel grapnel that is attached by a short line to the railings, and snags the pot line just below the buoy. When he occasionally misses, the urgency with which he retrieves the grapnel to recast it reflects the need to get on with the job—but more his self-annoyance. With buoy now in hand he persuades enough loose line towards him to run up over the winch block—slightly out from the gunwale and 2 metres above deck level—and makes a single loop around the winch drum. The skipper adjusts the position of the vessel so that it is more or less directly over the pot. A hand control starts the winch and second and third turns are established on the drum to give the rope traction. During hauling there is an almost magical balance between the crew's effortless drawing of the rope away from the winch drum on the one hand, and the full weight of the pot and its contents, and its resistance to being dragged up through the water column, on the other. The water is deep, close to 150 metres, and the pot takes a few minutes to haul. Visibility is at least 30 metres, and the first sign of it is a dark eerie apparition way below.

This is the exciting bit. Packhorse give the pot a reddish-brown appearance, which becomes more obvious as it nears the surface. The dark colour categorically indicates there are packhorse in the pots, but how many layers of them are there? The pot can be a third or more full if the lobsters are really on the bite. The pot breaks the surface and the crewman allows the line to slip on the winch drum as it hovers just above the deck railings, his hand steadying it. Then he pulls it towards him, at the same time releasing the rope. The pot lands on the deck with a crash.

The bait is gone, but not the lobsters—even though they could easily have vacated. Some pots are indeed one third full. The crew rush to separate out the pot's keepers from the other lobsters that are too small or otherwise illegal to land, and to re-bait the pot; with their heads down, they can have little inkling as to how soon they will be alongside the next float. The keepers go into live tanks to which fresh seawater is constantly supplied. Lobsters to be returned to the sea may not always be treated with quite the same care as the keepers. Maybe there is just a little sense that these lobsters no longer belong to this vessel, so who cares? After all, they may never be seen again or worse, they may wander into someone else's pot.

Packhorse are unusual in that pots set for them seldom contain any lobsters if hauled after just one night. Leaving the pots for at least three nights is the usual method. Even after a couple of weeks undisturbed, pots may contain dozens and dozens of packhorse, as if the lobsters are enjoying a convivial, sheltered break from their doings. So in these remote grounds of the Far North, many hours' steam from the nearest port, the hauling of individual pots is typically a once-in-a-while affair. This in stark contrast to the fishing practice for the red rock lobster *Jasus edwardsii* in most parts of New Zealand—and the majority of other species elsewhere—where pots are hauled daily. Spiny lobsters feed at night, so typically boats leave port near first light, and return in the early to mid afternoon to be met by the truck from the packing shed. The early start means that fewer of the lobsters will have left the pots before they are hauled; much of the pot-hauling and resetting is over and done with before the sea breeze comes to anything; and the lobsters can be handed over to the packing shed's care well before day's end. (It also means there is less reason for suspicion that someone else has got to your pots first.)

Commercial fishermen catch packhorse using pots—steel- or wooden-framed and covered with synthetic or steel mesh. For the Cape Reinga grounds the pots need to be particularly heavy because of the strong tidal currents and because they need to hold their position on the often sandy substrates during storms. Accordingly they are typically close to 2 metres by 2 metres and about 60 centimetres high, the extra steel welded to their base bringing their weight up to about 200 kilograms. (Most of the millions of spiny lobsters landed each year around the world are caught in pots much smaller and lighter than these.) There are two or three large necks on the top, each with either a steel-mesh or plastic funnel protruding into the pot that means it is easier for a lobster to enter than to leave. The bait—usually heads and frames (bodies which have had their fillets removed) of oily fishes such as kahawai—is suspended in the top of the pot away from both the sides and the funnels, or held within a hanging plastic mesh container or two. The pot is connected to the air-filled plastic buoy by synthetic rope 12 millimetres in diameter that can hold weights of well over a tonne. All up, a pot rigged for fishing these depths won't leave you with much change from $1000.

These large pots earn their keep once fishermen get onto runs of packhorse. With uncanny ability, a certain few individuals have over the years been able to find and track them—not the small packhorse as they migrate up the east coast of the North Island, but the larger, mature individuals as they move inshore and offshore each year in the Far North, in accordance with their natural cycles. No one really knows what packhorse look like as they move across the fishing grounds—how individual lobsters respond to their first whiff of the bait, how they might jostle as they crawl around, up and into the pots—but the numbers often found in the pots suggest the lobsters are more likely to be moving in swathes than in single-file.

The mesh-covered, steel-framed pots used for packhorse are larger than those used in most other spiny lobster fisheries. Because usually the lobsters can readily leave the pot, they are not really traps.

Working pot by pot we complete a size composition by sex—the carapace length measured to the nearest millimetre of every lobster—and tag the undersized ones. Size compositions collected over time from similar-style pots set in much the same place and at the same time of the year can be used to help pin down the status of a fishery. A pristine—unfished—population presents at first fishing a great variety of lobster sizes, certainly including the very large, but usually too a sprinkling of the smaller individuals. (Small lobsters may be underrepresented in the pots though, the larger ones excluding them from the bait.) But as time goes by and fishing takes its toll, fewer of the large lobsters remain and the mean size decreases.

Additional detail recorded for those tagged includes any damage such as missing limbs. Broken or missing appendages are replaced over successive moults, but at the expense of the body-length increment at each moult, something to take into account when analysing the growth of recaptured lobsters. Packhorse cope well with the loss of a limb: it is usually replaced to about two thirds of its original dimension at the first moult, and is difficult to distinguish as a regrowth after the second. This is why, using scissors, we amputate the outer third of one of the pleopods. Upon recapture, it will be straightforward to see if the lobster is unmoulted, has moulted just once (the original straight cut will show part regeneration), or has moulted twice or more (complete regeneration).

Occasionally the sides of the carapace are flexible to the touch and there may be a pink hue to the tail muscle when viewed from beneath. These lobsters are about to moult. Calcium and protein from the exoskeleton have been re-absorbed into

Measuring the carapace length of a small packhorse before tagging it.

the blood, leaving otherwise hard structures, particularly in the branchial region, thin and plastic. Break off the tip of a feeler and you are left with a limp portion of the developing new exoskeleton drooping from the end. Very seldom, however, is any lobster in a pot soft all over—as they are immediately after a moult. The open seafloor is no place for a recently moulted lobster because this is when they are most prone to predators. The soft or leathery appendages of a newly moulted lobster are easily removed one by one by even small predatory fish, leaving the lobster marooned and powerless to resist further attack.

'The Moray [an eel that dwells in holes in the reef] loathes the Octopus and the Octopus is the enemy of the Carabus [spiny lobster] and to the Moray the Carabus is most hostile,' wrote Aelian centuries ago.[1] These writings suggest that even in the Hellenistic-Roman era, the complex ecological relationships between the denizens of the depths were noted; they also confirm the universality of spiny lobsters not getting on at all well with octopuses. This is something lobster fishers everywhere will attest. Invariably the octopus has spooked and then dealt to the largest lobster in the pot before vanishing through the meshes, leaving the lobster's empty shell as its calling card. The method of this persistent and effective predator is to subdue the lobster with its tentacles and then to bite into its softer under-tail with its horny beak. Juices expressed into the flesh immediately begin the process of digestion. The smaller of the packhorse may meet their end by this means, but Adam Davey says that it is usually the other way round with large packhorse. Often in his pot would be a corpse, not of the packhorse but of the octopus, badly mutilated.

φφφ

Both the *Medea* and *Mako* were working the fabulous Far North packhorse grounds. How and by whom was this fishery—the one really serious packhorse fishery in New Zealand—discovered and developed? Other accumulations of large packhorse along the east coast of the North Island had been fished down to uneconomic levels within a few years of being found, but this one in the Far North had survived decades of fishing and hundreds of tonnes of extractions. Stories abounded of fishermen with catch rates so remarkable as to be almost unbelievable, but one name stood out above all others: Bill Hopkins.

Spiny lobsters came under the Quota Management System in 1990. The size of the individual transferable quotas (ITQs) fishermen received was based on their landings from 1982 to 1986. The average annual landings of each fisherman, based on the monthly fishing returns they had supplied the Ministry of Agriculture and Fisheries for those years, was the starting point for deciding their individual annual quotas. Bill Hopkins' quota was so short of what he thought it should have been that he asked the Ministry for copies of his original returns. It was then the source of the discrepancy became clear. For example, in the month of March 1983, using 60-odd pots, he had caught and landed a total of 29.6 tonnes of packhorse, in several individual landings.[2] He recorded his landings to the nearest kilogram even though the form allowed for them to be recorded to a tenth of a kilogram. After all, he was dealing literally with tonnes and tonnes. The column of figures in Bill's large flowing handwriting sometimes took into account the pre-printed decimal point; other times it didn't very well—there wasn't much room on the form. But the form had additional figures, in a different style and overwriting Bill's. Clearly a conscientious clerk at the Ministry office, faced with having to decide where the decimal point in the catch figures should lie, had thought it impossible for any fisherman to have caught several tonnes over just a few days—and so had altered all his catches by one decimal place. This reduced Bill's monthly catch to one tenth of what it had actually been.

Using landing dockets, payment slips and other documents, Bill was able to demonstrate the truth concerning his catches. His packhorse catches for the five criteria years were close to 28, 47, 19, 7 and 4 tonnes respectively. His final quota allocation was 23 tonnes, more than half of the eventual Total Allowable Commercial Catch (TACC) for packhorse.

Large and variable as these early- to mid-1980s catches were, they were modest compared with what Bill had previously taken. No one else, then or since, caught them in such quantity. Fishing out of Russell on FV *Carol Ann* in the early 1960s, he first explored as far south as Tolaga Bay, as well as to the north. And one area impressed him more than any other. His springtime scout pots on reef edges in Spirits Bay showed good colour, but catches on the surrounding sand were staggering. Many of the packhorse were huge egg-bearing females, but among them too were good numbers of similar-sized males, together with very commercial quantities of smaller lobsters. Ken Turner, trawling with the *Marine Star* along one of the few lines existing near Bill's pots, hooked up on several tonnes in one regretted tow: his badly damaged net took ages to retrieve. Clearly the stocks of packhorse in this area were phenomenal.

 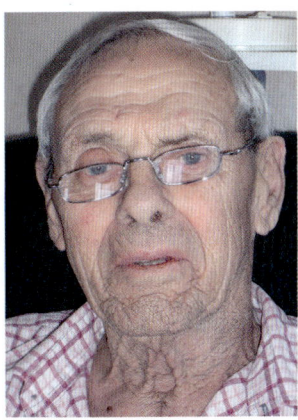

Bill Hopkins worked his vessels and crews mercilessly, becoming one of the most successful spiny lobster fishermen ever. Often the packhorse were held in holding pots (left, with Bill to the right) before being delivered to port. Bill Hopkins died at the age of 80, soon after this photograph of him (right) was taken.

Bill Hopkins summarised his time in the Far North packhorse fishery in correspondence with the Ministry and in interviews.[2] 'In the 1960s, I caught large quantities of packs in Spirits Bay in FV *Carol Ann* and later FV *Provider*. Lifts for one day peaked at 80 sacks or 4 tonnes.' It was not rare to land half a dozen tonnes after a two-day trip. 'It was a piss-poor day if there was only a tonne,' he told me. In April 1970 he discovered new packhorse grounds at depths of 90 metres off the west end of Spirits Bay, and later others 30–37 kilometres west and southwest of Cape Reinga at depths down to almost 200 metres.

His monthly catches often exceeded 10 tonnes. If handling this quantity of lobsters on board relatively small (15–20-metre) vessels was a challenge, getting them to the factory live—as the law required—was a remarkable feat. It was possible only because packhorse are particularly hardy. They were jammed tightly into jute sacks held open by hooks on a metal frame during filling, each sack holding 50 kilograms or so. Large packhorse were so difficult to bag, invariably hooking the tips of their legs into the fabric, that the crew got into the habit of leaving them for a while in the sun to quieten.

But by the late 1980s most of New Zealand's spiny lobsters were being exported live so it was essential that the packhorse were landed active and as intact as possible. Bill held his packhorse in live tanks either on the deck itself or within the hold, according to vessel, and had seawater pumped evenly among the lobsters through perforated piping. Deck tanks were drained temporarily ahead of heavy weather to maintain vessel stability.

Bill Hopkins retired from fishing in 2005. It's not possible to establish his entire catch record but what is clear is that he caught many hundreds of tonnes of packhorse, with some phenomenal daily catches. Graeme Eccles of Select Seafoods told me how, in the late 1970s:

> He [Bill Hopkins] left the port of Houhora about one pm one Saturday, and the next morning at seven thirty brought ashore at Scott Point [a little south of Cape Maria van Diemen] 44 sacks, each around 140 pounds, for me to pick up. These lobsters could have come from only one—or at most two—lifts of his pots! And he only had a few pots—but big buggers.[3]

Most seasons his catch was in the tens of tonnes, frequently around 30, and peaking at 112 tonnes in 1970–71 (which included 33 tonnes in just one month). At present values, 112 tonnes would be worth about $4 million—back then he received $120 000.

These catches may seem over the top from a present-day perspective. A catch today of even 20 tonnes by a single vessel in one year is extraordinary. They were, nevertheless, taken within the regulations of the time—and not without great effort and, at times, risk. There is every chance that Bill's pot catch rates by weight were the highest that will ever be seen for any spiny lobster, anywhere. It is also very likely that he landed the largest packhorse ever taken in New Zealand waters, and probably, therefore, the biggest spiny lobster ever caught on Earth. This is a record unlikely to be matched, given that Bill Hopkins fished packhorse in the Far North from the very beginning of its commercial exploitation there, which is when the most real giants would have been around.

MAORI TRADITIONAL FISHING

The fishery was at its zenith when Bill Hopkins made his catches, yet inshore stocks of spiny lobsters had been exploited by Māori for centuries. Although the remains of spiny lobsters usually decay rapidly, their mandibles survive some environments, and middens show that lobsters were a significant food for Māori well before the early European arrivals. Spiny lobsters have turned up in excavations in Palliser Bay at the southern end of the North Island, in Fiordland in the southwest of the South Island, and at the Chatham Islands east of the South Island.[4] The evidence from Palliser Bay is that Māori collected all sizes of lobster, not preferentially targeting the larger ones.[5] The fishing pressure was such that it significantly reduced the mean size of the lobsters. But there is also suggestion of rotational harvesting, with areas remaining unfished for long periods. This points to quite different protocols to those which pertain today in most spiny lobster fisheries around the world.

These midden excavations—all undertaken well beyond the area of main occurrence of packhorse—revealed only red rock lobsters, but the lobsters in other reports may well have included packhorse. Natural history illustrator Sydney Parkinson, on Captain James Cook's first visit to New Zealand, wrote of Tolaga Bay: 'This bay abounds in a variety of fish, particularly shell and cray-fish; some of the latter, which we caught, weighed eleven pounds; these are found in great plenty, and seemed to be the principle food of the inhabitants, at this season of the year . . .' Archaeologists Foss Leach and Atholl Anderson were to later speculate how the larger of these lobsters may well have been *Sagmariasus verreauxi*.[4] In fact spiny lobsters featured many times in the observations of the early visitors. James Cook himself wrote 'These we also bought everywhere to the northward in great quantities of the natives, who catch them by diving near the shore and finding out where they lie with their feet.'[6] A few years later, on James Cook's second visit, natural historian George Forster was to write of sailors trading for dried 'craw-fish', which quickly became an important item of barter with the whaling and sealing ships that arrived soon after the first European explorers.[6]

The manner in which Māori on the east coast of the North Island harvested their spiny lobsters in the latter part of the 19th century was described by ethnographer Elsdon Best in his *Fishing methods and devices of the Maori*.[7] Although taruke

Joseph Banks trading a piece of tapa cloth for a red rock lobster during James Cook's first voyage to New Zealand, on *Endeavour*, probably sketched by the Polynesian navigator Tupaia.

(pots) were the most widely used method, both men and women took lobsters by hand—ruku koura. They always descended feet-first rather than diving head-first. Men stripped; women wore a maro (apron). A man who was energetic, daring, and successful at this pursuit would be referred to as toa ruku koura. But women were often the more expert, descending to depths of almost 6 metres.

The early 20th century anthropologist Peter Buck, under his Māori name Te Rangi Hiroa, provided us with more detail of how Māori of the Gisborne and Bay of Plenty regions went about harvesting spiny lobsters.[8] There were four methods. In addition to diving and potting, they used matire and pouraka. Matire was a simple method which used a rod with a number of strands of flax over one end. Amidst the strands was the bait, a pāua (abalone). The rod was thrust down into rocky crevices and the lobster, attracted to the bait, got its legs tangled in the strands. A pouraka was a bag net with a hoop, handle, sinkers, bait attachment, and line. The net was about a metre in diameter, made of strips of flax, and the hoop at the mouth of the net was made of supplejack. Usually baited with pāua, it was lowered into the selected spot, the attached sinkers taking it to the bottom. In shallow waters, half a dozen pouraka might be used in rotation, the density of the lobsters being such that by the time the fisherman had lifted the last net, the first was ready to lift again. In deeper waters, poito (floats) made of light timber were used to mark the nets.

Red rock lobster pot or taruke koura (above);[7] and bag-net or pouraka, baited with a crab (right).[8]

Spiny lobsters close to the moult and soft on their sides—koura maunu—were most esteemed. At this time they were known to aggregate in certain spots (rua maunu, or simply rua koura), so precious as to be given names.[8]

Te Rangi Hiroa described processing the great delicacy koura mara:[8]

> This is prepared by soaking the crayfish in fresh water for about three days if the water is warm, and four or five if it is colder. The test is the loosening of the shell. When it comes away easily it is termed mahiti. People used to crayfish will then eat them raw and enjoy them: the smell is worse than the taste. When thoroughly mara the fish separates into three parts—the tuke (sternum), papa (upper part of the carapace), and hiku (tail). The flesh of the legs easily separates and comes away with the tuke. The flesh is placed on a wooden platform and support and left to dry for a day. Two tuke are placed together (karapiti), beaten or pounded to stick together, and exposed for another day. They are again beaten, cooked in an earth oven, and dried. When dry they are packed in baskets, and will keep for a year. The other parts are dealt with in a similar manner. The papa part is usually consumed by the family, but the tuke and hiku parts are kept in the storehouse for occasions. Crayfish preserved in this way, whilst very palatable, create a great thirst.

More recent Māori exploitation of spiny lobsters brings—not unexpectedly—a meld of the traditional and the contemporary. And a touch of melancholy.

Honourable Dover Samuels, Minister of Māori Affairs from 1999 to 2002, comes from near Matauri Bay, just inshore of the Cavalli Islands and a little north of the Bay of Islands. His hapū (clan) Ngati Kura is part of the Ngāpuhi iwi (tribe), whose best-known member was perhaps the chief Hongi Hika. Hongi Hika travelled to England in 1820, with the missionary Thomas Kendall and the chief Waikato, to meet King George IV. (While there he helped compile a Māori dictionary for the Church Missionary Society.) Dover's idyllic home is Putataua Bay, fringed with sedimentary reef and arching pohutukawa. It was here that the missionary Samuel Marsden met Hongi Hika shortly before reaching Rangihoua in the Bay of Islands to preach the first Christian service in New Zealand, on Christmas Day 1814.

Dover Samuels had a good-sized packhorse on display in his parliamentary office, but the size of this one pales in comparison to the 34-pounder he caught at the Cavalli islands.

'There is something truly majestic about packhorse,' Dover enthused. 'Their sheer size, strength, and aggression. Anyone would be in awe of them, particularly of the big ones. But these aren't big,' he said, pointing to the mounts on the wall. 'The largest I ever caught was 34 pounds. I was with Wade Doak at the Cavallis and swimming over an area of cobble and weed between reefs. All of a sudden the section of weed below me started to move. I struggled it to the surface, as it ripped my wet suit. But I wasn't going to let it go. It was a big old bugger, with weed and growths on its shell.'

Dover gave me a glimpse into the fishing of peke-hoiho (the transliteraton of packhorse—they are more traditionally known in Māori as pawharu) in the area.[9] In the early to mid-1900s there were 20 or so whānau (families) whose responsibility it was to provide koura (spiny lobsters) for the larger community at Matauri, plus some for sale or trade. (Other whānau provided different goods and services.) Typically their vessels were double-ended, whaleboat style, under oar and sail. Each vessel carried half a dozen pots made of supplejack, elongate and like a short hinaki (eel trap), but with the opening at the top rather than at the ends. A sisal rope was attached to the karewarewa (float), which consisted of metre-long poles lashed together at their centres. The junction was secured with many turns of harakeke (flax), giving further mass and flotation.

But packhorse at that time very much played second fiddle to the red rock lobsters. Packhorse were generally thought to have coarse flesh and were worth far less on the commercial market. What's more they played havoc with the pots, often breaking into them to get at the bait and at the other, more highly prized, pot contents—red rock lobsters and red moki (an oily fish, for smoking). So, as a boy, when packhorse catches exceeded requirements, Dover's task was to dispatch them with a wooden club. Nevertheless packhorse were an abundant source of meat when hosting large numbers of visitors at hui (gatherings) and tangi (funerals). A favourite way to prepare them—and several other food items such as corn—was as that semi-decomposed koura mara described earlier.

Dover Samuels commercially fished packhorse and reds out of Matauri Bay at various times between the 1960s and the 1980s. He easily managed the monthly catch-reporting requirements of the authorities, but many of the whānau did not; some of the older ones didn't write. And then, too, there were larger boats from the Bay of Islands and Whangaroa to contend with. Rather than resist, many whānau simply stopped commercial fishing under these unfamiliar terms. In so doing they lost access to a property right that would later have been theirs.

Today not one commercial spiny lobster vessel fishes out of Matauri Bay: there are still lobsters there but the fishing of them these days is big business, using mainly large vessels that can't be launched off a beach. But over summer and autumn the campground is host to many small boats, often with crews firmly focused on at least seeing—if not catching—one of the huge packhorse that are still to be encountered from time to time at the Cavalli Islands.

Both species of spiny lobster remain essential customary harvests of Māori, but methods of capture have changed. Free diving (now almost always head-first I believe) has been augmented with scuba, and pots are now much more often made of steel with a synthetic mesh cover. Māori are also major commercial fishers (and quota holders) of spiny lobster—as well as many other marine species.

COMMERCIAL FISHERIES

Brief as it has been, it is important to characterise and document the commercial fishery for packhorse as best we can before its detail is entirely extinguished from memory. For it is easy to suffer the historical myopia that is manifested in what Daniel Pauly of the University of British Columbia describes as the 'shifting baseline syndrome'.[10] This is where researchers and managers, in particular, base their understanding of healthy fish populations on what fish stocks are like during their lifetimes, with no framework for incorporating how much more plentiful fish may have been in past generations. An accurate and complete catch and effort record, right from the start of fishing of any particular stock, is what is required—but it rarely exists. The data for packhorse up to the late 1970s are incomplete, unreliable, and frequently a mix of two species.

The extent to which packhorse were part of the pre–World War II fishery is far from clear. It's hard to believe that they weren't fished commercially at all. Indeed, one of the few regional synopses of spiny lobster fishing of those times, the Report

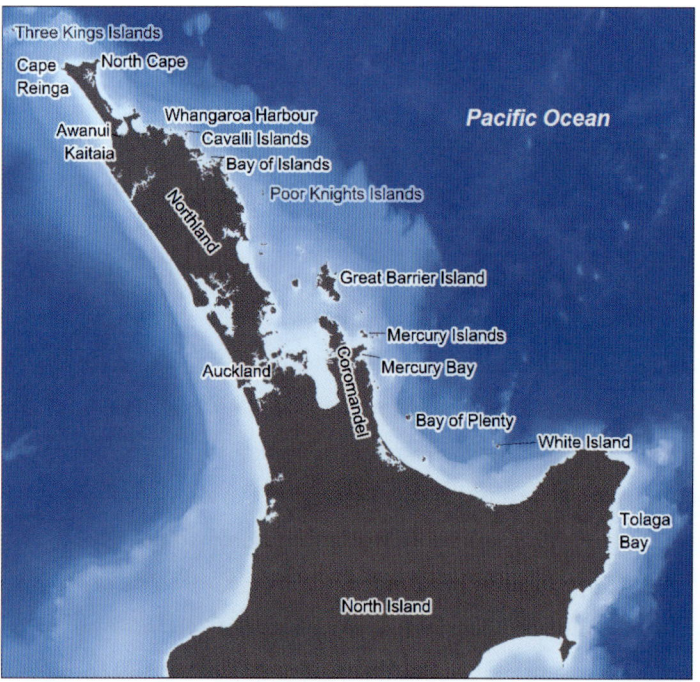

Places in northern New Zealand associated with the packhorse fishery.

of the 1937–38 Sea Fisheries Investigation Committee, indirectly referred to the existence of packhorse fisheries.[11] The report stated that, in Mercury Bay in the mid to late 1930s, 'The crayfish-men follow the crayfish out when they leave the inshore rocks and migrate to the sand and kelp bottom. When fishing under these conditions, they suffer damage to their gear by the action of the seine boats shooting their lines around the area where the pots are placed.' This is hardly an account of fishing for mainly reef-bound red rock lobsters. And it is probably no coincidence that, soon after the war, Mercury Bay was extremely important for packhorse, with precisely the same complaint voiced by packhorse fishermen.[12] The report goes on to say of Mercury Bay that there were 'heavy abstractions of up to 25 tons per week during the short time the export trade flourished'. And further north, the main grounds fished were from Matauri Bay (inshore of the Cavalli Islands, which themselves later yielded large catches of packhorse) to Whangaroa: 'When the canning factory was in operation up to 6 tons per week were being taken from these grounds.'[11]

It was the developing and seemingly insatiable market for frozen tails in the United States soon after the war that really lent impetus to spiny lobster fishing in New Zealand. And the progress of the packhorse fishery was essentially one of serial depletion, eventually leaving only one significant productive ground—the Far North. Newly discovered, usually small, pristine patches along the east coast of the North Island were intensively fished, with phenomenally high initial catch rates. These were followed by rapid declines and virtual extinguishment—all usually within a few short years. This pattern, common among spiny lobster fisheries worldwide, was even more dramatic for packhorse because there were two distinct components to the fishery. First were the smaller, northward-migrating juveniles, which in 1969 became illegal to land because they were too small under the new minimum legal size limit. Second were the populations of larger packhorse—individuals that for one reason or another had not migrated north, slowly accumulating locally over eons. These were never going to sustain heavy fishing pressure because of their extremely low rates of replenishment, so they virtually disappeared.

By the mid-1950s, a trawl and pot fishery for packhorse was well underway around the Mercury Islands and in Mercury Bay, particularly at the Hole in the Wall and Old Man Rock. The region was so productive that it became, not only *the* hot spot of the packhorse fishery, but also one of the most important spiny lobster fishing areas in the entire country. Apparently, the main fishery at White Island, the southernmost locality for any quantity of large, mature packhorse, also began in the 1950s, and was still productive in the early 1960s. Of White Island, John and Melody Anderson wrote:

> In the 1960s . . . an expedition from my dive club happened upon a march of these green monsters. Everywhere the divers looked, Packhorse crayfish were to be seen right out in the open. Every cylinder was sucked dry and every sack filled. One, a great chaff sack contained 16 crays and weighed over 80 kilos—it took three divers to get it back to the boat.[13]

Other places that provided large catches of packhorse for a few years around this time to were, from north to south, the Cavalli Islands, Poor Knights Islands, and the east coast of Great Barrier Island.[14] It has been said that packhorse catches around the Barrier were so large, and the island so far away from the packing houses, that many of these great old lobsters ended up at the dump.

A 1970 analysis of packhorse landings by the Marine Department's Craig Kensler and Walter Skrzynski showed how the centre of the fishery shifted from the Bay of Plenty to the Far North during 1962–66.[15] The North Cape Shortie

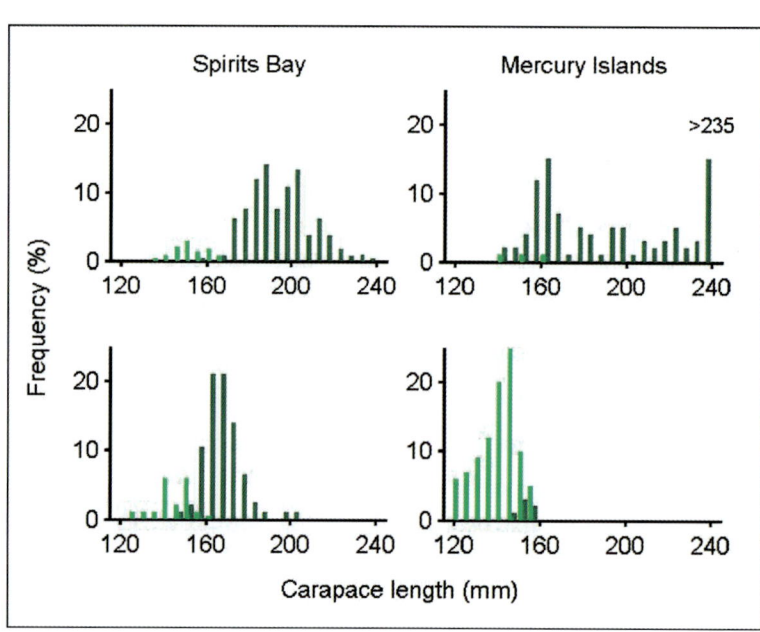

Size of female packhorse in typical spring-time catches from Spirits Bay and, further south, near the Mercury Islands in the western Bay of Plenty in the 1960s—near the beginning of each of these fisheries (top row). Catch samples taken in spring during the 1980s (bottom row) illustrate how fishing had reduced packhorse mean size in the north, and had removed almost all large individuals in the south, in just 20 years. Light green shows females that were without setae and therefore immature; dark green shows females with setae, the larger of which would have been mature.

Patch was first commercially potted in September 1961, and by 1966 about 10 boats fished there. It was also around 1961 that Bill Hopkins came upon the Spirits Bay stocks of packhorse. A few years later he revealed the deeper patches off Cape Reinga. There was a small amount of commercial trawling of packhorse in Spirits Bay in the early years, until it became clear it was more economic to pot them.[16] Vessel numbers in the Far North (including at North Cape) appear to have peaked at about 42 in 1971–72. The TACC of 40.3 tonnes came into force in 1990.

By the late 1980s almost all packhorse were being exported live by air, mainly to Hong Kong and China. At Awanui's Select Seafoods, Graeme Eccles would 'swim' the lobsters for 3–4 days at 9°C in large tanks of recirculating seawater, during which time they would rid themselves of food wastes. Their metabolic

Estimated annual New Zealand commercial landings (whole weight) of packhorse, 1953–2008

Year	Catch	Year	Catch	Year	Catch	Year	Catch	Year	Catch	Year	Catch	Year	Catch
1953	*20*	1961	*80*	1969	*80*	1977	*160*	1985	20.3	1993	5.7	2001	7.8
1954	*20*	1962	*120*	1970	*200*	1978	*60*	1986	7.7	1994	7.9	2002	8.6
1955	*40*	1963	*120*	1971	*200*	1979	29.1	1987	10.2	1995	23.8	2003	16.4
1956	*40*	1964	*120*	1972	*100*	1980	11.2	1988	15.0	1996	16.9	2004	20.8
1957	*40*	1965	*120*	1973	*100*	1981	25.2	1989	10.0	1997	16.2	2005	25.0
1958	*40*	1966	*120*	1974	*100*	1982	28.9	1990	1.8	1998	16.2	2006	25.4
1959	*40*	1967	*80*	1975	*100*	1983	73.1	1991	21.2	1999	12.6	2007	34.1
1960	*40*	1968	*80*	1976	*100*	1984	36.8	1992	15.3	2000	9.8	2008	36.3

Once delivering annual quantities in the (low) hundreds of tonnes, the packhorse fishery is now tiny, both globally and in the New Zealand context. But nevertheless it remains locally important. The table gives estimated annual commercial whole weight landings in tonnes, based on official figures, publications,[17,18] credit notes from processors, extensive interviews, and other documentation,[14] measured in calendar years until 1978 and fishing years from then on. Landings before 1953 are too poorly known to be included. Years for which the estimated landings did not concur with official figures, or there was no official value, are shown in italics. Virtually all landings before 1961 were from the east coast of the North Island south of North Cape, those for 1961–62 from both the east coast and the Far North, and from 1962 onwards from the Far North. The Total Allowable Commercial Catch of 40.3 tonnes, introduced in 1990, has never been caught.

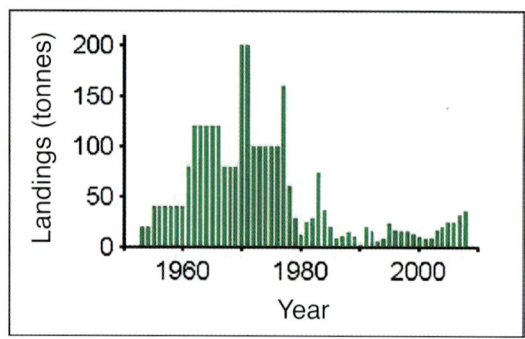

Estimated annual New Zealand commercial landings (whole weight) of packhorse, 1953–2008. Customary and recreational catches are unknown but are probably tiny relative to the quantities caught by commercial fishers.

rates fell at the same time. For the final 12 hours the water was taken down to 3–4°C; the lobsters were then packed in polystyrene boxes. A couple of containers of frozen seawater were added to each box to help maintain low temperatures on the way to market. The trip, from tank to tank, would normally be less than 24 hours, but the packhorse were capable of surviving at least 40 hours out of water provided temperatures remained low.[3]

After well over half a century of commercial packhorse fishing in New Zealand, the Far North is the only area that continues to yield legal-sized packhorse in any quantity. It was most productive between 1962 and 1978, peaking in 1970–71 with catches of around 200 tonnes each year. Over the past three decades, annual landings have fluctuated without pattern, most often between 10 and 20 tonnes, until very recently. More focused fishing in the past decade has led to landings that are now approaching the TACC for packhorse.

φφφ

Recall, though, that *Sagmariasus verreauxi* is not fished only in New Zealand. The southeast coast of Australia—particularly New South Wales, where the species is known as the eastern rock lobster—is the other mainland region where it is found. And the commercial fishery for *S. verreauxi* in New South Wales is a much larger affair than it is in New Zealand, and has been going a lot longer—since at least the 1880s. This valuable little fishery supplies a high-value 'boutique' market, including well-known seafood restaurants on the Sydney waterfront. Little is exported. The minimum legal size—104 millimetres carapace length—is much smaller than that in New Zealand (where for females it is equivalent to 155 millimetres carapace length). There is also a maximum legal size—180 millimetres carapace length—which helps look after the stock of large breeders. Scott Westley, who fishes out of Jervis Bay, 200 kilometres south of Sydney, and holds 8 per cent of the commercial quota, helped sketch the following account of the way the fishery runs.[19]

A TACC of 128 tonnes limits extractions; that amount is fished by about 80 vessels altogether but just a dozen or so take the lion's share. Each new fishing year starts on 1 July (although fishing takes place all year round), and there are two distinct components to the commercial fishery. The inshore fishery, from early in the fishing year until late December, involves smallish, mainly beehive pots set on the reef and surrounding sand in waters down to 15 metres deep. It is aimed at the smaller lobsters that have just reached legal size, each weighing about 0.7 kilograms. The offshore fishery, in waters 100–200 metres deep and up to 20 kilometres from shore, is only for larger vessels using large pots on open mud and sand bottoms, from January to June. Here most lobsters harvested are about 1 kilogram. Most of Scott's catches are from these offshore stocks—and are thought to be lobsters migrating to join the spawning stock in the north.

This is a fishery which has been hammered but is now rebuilding. Catch per unit effort doubled between 2000–02 and 2006–08, and the commercial catch is now constrained by the TACC. The size of the breeding stock has improved markedly and now approaches the target 30 per cent of its unfished level. At a low ebb until recently, the New South Wales eastern rock lobster fishery now seems to be coming right.

φφφ

Although locally important, the packhorse fisheries in both New Zealand and Australia are trifling compared with the world's larger spiny lobster enterprises. And given their unit value, it is little wonder that spiny lobsters support some of the most lucrative of all fisheries. We've all seen them in the fish market and flinched at the cost. They are often priced per 100 grams, as if to assuage the blow. But you are up against what the most discerning and wealthy diners from Brussels to Beijing are prepared to pay. The most important of these spiny lobster fisheries are highly specialised and single-species. But in other places the lobsters are taken incidentally or as part of mixed-species fisheries, and they also sustain numerous, far-flung artisanal enterprises.

Large industrialised spiny lobster fisheries are fairly recent, most of them—as we saw with packhorse—only really getting up to speed after World War II. Up until then fishing had more often than not been at a small scale for local consumption. More efficient refrigeration and burgeoning air freight services—and, particularly, a very healthy market in the United States—changed all that. World spiny lobster landings built rapidly, and annual production had reached around 70 000 tonnes by the 1970s.

All the main shallow-water and near-shore spiny lobster fisheries are fully exploited, any occasional new areas usually being small outlying pockets that are at first fantastically productive but which are rapidly fished down. The real biggies belong to the genus *Panulirus*. In 2005, the Caribbean spiny lobster *Panulirus argus* (principally in Cuba, Brazil, the United States and Mexico) and the western rock lobster *Panulirus cygnus* (Western Australia) together made up 60 per cent of the world's total spiny lobster catch of just over 80 000 tonnes, the Caribbean spiny lobster making up the lion's share.[20] Other *Panulirus* species brought this proportion up to 85 per cent. Next most important were the *Jasus* species, mainly the southern/red rock lobster *Jasus edwardsii* in Australia and New Zealand and the Cape rock lobster *Jasus lalandii* in South Africa, contributing 13 per cent of the total catch.

Despite its common name, the Caribbean spiny lobster is found widely in the west North Atlantic, from North Carolina (United States) to just south of the Equator.[21] It lives on sand and rock substrates down to 100 metres. This lobster is commercially fished by more than 20 nations, at both large industrial as well as artisanal scales, mainly using pots, bully nets (nets attached to long handles and used in shallow waters), and hand-gathering by divers. The largest national landings in 2005 came from the Bahamas (9000 tonnes), Brazil (7000 tonnes), Cuba (6000 tonnes), Nicaragua (4000 tonnes), and Florida in the United States (1500 tonnes).[20] In several regions, artificial habitats are used to concentrate lobsters, and they may also enhance production.

The fishery for the western rock lobster runs along the coast of Western Australia from Shark Bay south. There are about 550 boats fishing 57 000 pots focused on catching juveniles and young adults that have just reached the minimum legal size

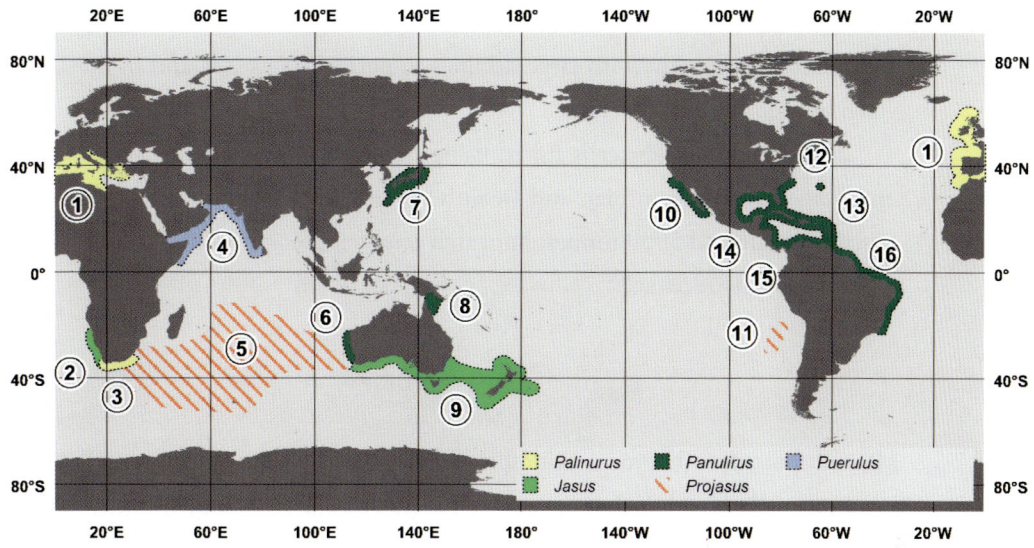

World distribution of the main spiny lobster fisheries.

Patches of colour denote the main spiny lobster fisheries, but spiny lobsters are exploited much more widely than this when you include bycatch from other fishing methods such as trawling, and small-scale commercial and artisanal fishing. (1) the European spiny lobster *Palinurus elephas*; (2) the Cape rock lobster *Jasus lalandii*; (3) the southern spiny lobster *Palinurus gilchristi*; (4) the Arabian whip lobster *Puerulus sewelli*; (5) the Cape jagged lobster *Projasus parkeri*, in certain parts of the area shown; (6) the western rock lobster *Panulirus cygnus*; (7) the Japanese spiny lobster *Panulirus japonicus*; (8) the ornate spiny lobster *Panulirus ornatus*; (9) the southern/red rock lobster *Jasus edwardsii*; (10) the California spiny lobster *Panulirus interruptus*; (11) the Chilean jagged lobster *Projasus bahamondei*; and (12–16) the Caribbean spiny lobster *Panulirus argus* in Florida, Cuba, Mexico, Central America and (with the smoothtail spiny lobster *Panulirus laevicauda*) Brazil, respectively.

and which are about 4 years old.[22] This fishery was awarded Marine Stewardship Council certification in 2000 as a well-managed fishery, the first in the world for a spiny lobster. Annual landings have, until recently, been fairly constant at around 11 500 tonnes, fluctuating about 20 per cent around the mean and in line with the strength of the previous puerulus settlement. The disturbing change that has recently taken place in this fishery is discussed in the next chapter.

The fishery for *Jasus edwardsii* takes place along the south and southeast coasts of Australia (southern rock lobster), and around mainland New Zealand and at the Chatham Islands, 800 kilometres east of the South Island (red rock lobster).[23] Landings in 2005 totalled 7000 tonnes, two-thirds from Australian waters. The only significant commercial fishing method is potting, to depths of 300 metres, but mainly at less than 100 metres. In most places, the majority of the lobsters taken are new recruits to the fishery. Annual landings throughout have tended to fluctuate significantly, with large declines in parts of New Zealand in the 1980s having been recently reversed.

The Cape rock lobster is fished using hoop nets and pots to depths of 100 metres along the Namibian and (mainly) South African coasts. Annual catches used to be around 5000 tonnes, but catch rates fell precipitously about 1990 and are still to fully recover.

φφφ

Baited pots are the most widespread method used to catch spiny lobsters commercially, but there is quite a wide variety of other fishing methods employed around the world. These include spearing, trawling, netting (drag, tangle or trammel, gill, hoop, set, seine, hand, bully), taking by hand while diving or searching reef tops, and using baited lines. Those most in need of explanation are as follows.

Tangle nets are used particularly in the Mediterranean Sea and east North Atlantic Ocean to catch the European spiny lobster *Palinurus elephas*, and in Asia to take the Japanese spiny lobster *Panulirus japonicus*. When lobsters come in contact with the net, either through walking into it or after being attracted to it because of previously enmeshed fish, they become ensnared.[24] The nets are set in strings, or sets, about 2 kilometres long, and the fishing vessels typically lift about 100 kilometres of them each month.[25] Generally they are left in the water for 3 days before hauling, to allow lobsters to be attracted to the fish caught in them. The host of black marks against tangle netting lobsters includes indiscriminate catching of individuals that cannot legally be retained, damage to and mortality of lobsters, and collateral damage to structure-forming seabed creatures. Their use has been implicated in the overfishing of the European spiny lobster, whose catch numbers declined from almost 9000 tonnes in 1988 to half that in 1996—with no evidence of recovery.[24]

Several spiny lobster stocks were trawled in great quantity soon after discovery. For example, during the 1960s up to 5 tonnes per trawl of the Natal spiny lobster *Palinurus delagoae* were taken off southern Africa.[24] Most trawling of spiny lobsters is now, thankfully, illegal and belongs to the past—as a fishing method it is indiscriminate, damaging to seabed communities, and often wasteful. But several species of lobster continue to be taken as bycatch by trawlers targeting other species, and it is conceded that trawling remains the most practical method to harvest certain (mainly deepwater) species.

Diving, sometimes with underwater breathing gear (hookah—where the diver is supplied with air from the boat above—or scuba), is the commercial fishing method most commonly used for the many tropical *Panulirus* species that do not routinely enter pots. There is, however, only one significant com-

Packhorse were briefly trawled in the Far North in the 1960s.

mercial spiny lobster fishery where lobsters are harvested by hand on natural habitat. Native islanders free-dive or use hookah from outboard-powered dinghies to take the ornate spiny lobster *Panulirus ornatus* in Torres Strait.[22] Annual catches are around 500 tonnes.

One other method of harvesting spiny lobsters merits attention. The Caribbean spiny lobster in Cuba[26] and Mexico[27] is caught mainly by means of artificial shelters (called pesqueros and casitas cubanas, respectively). Made of lengths of palm, sheets of fibro-cement or covered car tyres, most have a surface area of about 4 square metres. They are set on open sand bottoms and provide a large flat crevice in which lobsters take refuge. Up to 200 individuals are caught each night in a shelter. In recent years, pesqueros have been by far the most widely used item of fishing gear in Cuban waters, taking fully half of the 6000–10 000-tonne annual landings. Raúl Cruz and Bruce Phillips described this method of fishing in Cuba like this: The main lobster grounds are shallow, very clear and calm areas where the seabed is covered in seagrass. Fishermen place their pesqueros in groups of 15–20 on a zigzag line, with about 25–30 metres between each pesquero. A wooden bucket with a glass bottom allows the fisherman to detect lobsters in the pesquero. A net (chinchono) is placed around the pesquero and a prickle—a stick

Tangle nets are used to catch the Japanese spiny lobster and other closely related species in Asia (left). These Arabian whip lobsters, a deep-water species that is both target and by-catch for trawlers in parts of the Arabian Sea and Gulf of Oman, appear remarkably undamaged (right).

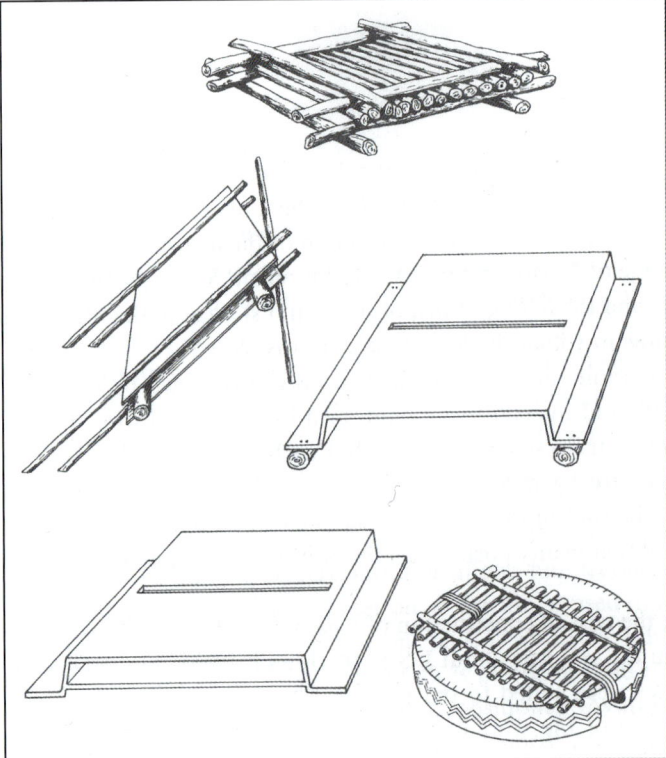

Pesqueros used in the Cuban fishery for the Caribbean spiny lobster.[26] Similar devices (casita cubanas) are used on the Caribbean coast of Mexico.

with an L-shaped wire at one end—is used to persuade the lobsters out of the shelter into the cod end of the net. Alternatively the shelter is tilted onto one side to get the lobsters to flee into the net.[26]

But as Patricia Briones-Fourzán and her colleagues emphasise, the use of artificial reefs to harvest lobsters is controversial.[27] It is still is not known whether they provide extra shelter that increases the carrying capacity of the area and eventually the biomass of reef-associated organisms including lobsters (the 'enhancement' or 'production' hypothesis), or whether they merely attract and aggregate organisms from surrounding areas without increasing total biomass (the 'attraction' hypothesis). If the enhancement hypothesis holds true, then the shelters may provide a powerful means to mitigate the impacts of habitat loss and overfishing. They may also serve as an effective management tool to increase carrying capacity and achieve sustainable resource use. The attraction hypothesis, on the other hand, implies that the use of artificial shelters could lead to overexploitation of the resource and the eventual collapse of fisheries.

RECREATIONAL FISHERIES

Spiny lobsters are also the basis of popular recreational fisheries. Packhorse, particularly large ones, are keenly sought by recreational divers in northern New Zealand: for the table; as trophy catches to mount and display; for evidence of dive prowess; and, increasingly, simply to observe (and maybe photograph) in situ. From time to time photos (often old, grainy black and whites) and anecdotes turn up that mesmerise. 'The two largest were 2 feet 6 inches from the horn between the eyes and the tip of the tail. They were huddled around a piece of kelp on sandy bottom,' wrote John Coombes of packhorse taken while diving for scallops in Opito Bay, a little north of Whitianga, around Christmas 1975.[29] The accompanying photo is of the legs of two men (well, the legs are hairy) who are each holding two packs by the bases of their feelers; if the tails weren't partly curled up, they would be stroking the ground. And the occasional large packhorse is still to be taken to this day off the east coast, in areas now bereft of any significant commercial catch of the species. Presumably these are lobsters that had not migrated north, and had managed for years to avoid capture.

John Coombes' 1975 dive catch near Whitianga, in Mercury Bay.[29]

The recreational size limit for packhorse is the same as in the commercial fishery—216 millimetres tail length. Most are harvested using scuba, but pots are also used. It is illegal to spear them (mainly because they could be undersized), or to take those soft-shelled (mainly because they cannot be accurately measured) or bearing eggs. The daily bag limit is six—awfully generous considering the size of this lobster.

Recreational fishers in most countries are permitted to harvest a limited number of spiny lobsters—typically between 2 and 12 each day—for their own use.[28] (Notable exceptions are Brazil, Mexico and Japan, where spiny lobster fishing is restricted to commercial fishers.) But there is little other consistency in the manner in which recreational fisheries are managed. Snorkel or free diving is the most widespread method of harvesting, but in many places scuba and hookah—and pots and/or drop nets—are also allowed. Queensland, Australia, is one of the few places in the world where recreational divers are allowed to use spears and spear guns to harvest lobsters.

Recreational fisheries can be a significant part of the total spiny lobster catch in an area, yet reliable estimates of the harvest—particularly if they are to be for more than just a few small segments of the fishery—remain elusive. Comprehensive data are generally available only in those few jurisdictions in which fishers require a license. And there is understandable reluctance among politicians to compel fishers to purchase licenses to harvest lobsters (which are often argued to be common property) for personal consumption. It is no coincidence that the parts of the world with the most useful long-term recreational catch data are those with the longest history of recreational fishers being licenced.

The effort—and catch—in what is probably the most intensive recreational spiny lobster fishery anywhere, for the Caribbean spiny lobster *Panulirus argus* in Florida, is astonishing. More than 100 000 fishing permits are issued annually.[31] Recreational fishers can't use pots. Instead they catch their daily bag limits of 6–12 lobsters (depending on their location in Florida) by diving and by using bully nets. The recreational catch is 30–40 per cent of the size of the commercial catch, with interest intensely focused over a couple of short periods of the year. First, there is the 'Special Two-Day Sport Season' just before the opening of the commercial fishery, during which up to 250 tonnes or so of lobsters may be taken. Then there is the first month of the regular recreational season, coinciding with the first month of the commercial season, when up to 800 tonnes of lobsters are extracted by recreational fishers.

Another very large recreational fishery is for the western rock lobster *Panulirus cygnus*. It accounts for about 600 tonnes of the landings each year from the shores of Western Australia,[28] although this figure pales against the annual commercial catch of around 11 000 tonnes.

There are no reliable estimates of the recreational harvest of packhorse in New Zealand. Probably the actual numbers are quite small—certainly compared with the numbers of red rock lobsters taken. Because of their size though, the total weight of packhorse taken recreationally each year may not be insignificant. And recreational fishing of this species in Australia is very popular: around 60 000 people participate each year, taking 26 tonnes.[32]

Brian Hoult caught this 6.3-kilogram packhorse on a hook while snapper fishing in Bream Bay in 2003.[30]

CHAPTER 8

MANAGING THE FORTUNES OF FISHERIES

'These contrasting situations offer an excellent lesson on the need to adopt a risk-averse approach to fishery management, and hopefully avert fishery collapse, which sometimes results from the joint effects of overexploitation and poor environmental conditions.' So wrote North Americans Rom Lipcius and David Eggleston in 2000 about two *Jasus* fisheries—one in New Zealand and the other in Namibia—with what were apparently very different management approaches.[1] But as it turned out, the main lesson from their examples was the importance of natural processes affecting spiny lobster abundance.

In New Zealand, the red rock lobster *Jasus edwardsii* fishery in Gisborne had been subject to a comprehensive management scheme, which was declared to be the cause of a marked increase in catch per unit effort (CPUE) in the mid-1990s. Scientists enthusiastically proclaimed it to be 'a fisheries management success story'.[2] In contrast, poor management, together with episodes of weakly oxygenated waters, had apparently led to frightful catch reductions in the Namibian fishery for the Cape rock lobster *Jasus lalandii* from the mid-1960s.

Lipcius and Eggleston's intent had been fair—but the insight misleading. Soon after their chapter had been published, the Gisborne (CRA 3) fishery, with its package of management measures well embedded, peaked—and then subsided at a rate never seen before in the fishery. Total landings and catch per pot

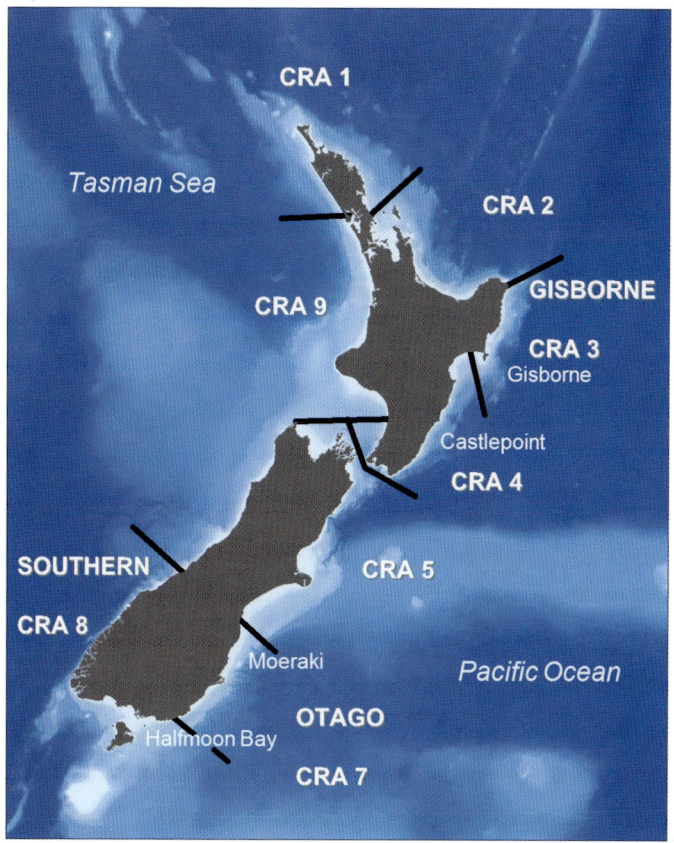

The commercial fishery for the red rock lobster *Jasus edwardsii* in New Zealand is managed as 10 fishery areas (each labelled 'CRA', short for 'crayfish'). CRA 6 is the Chatham Islands; CRA 10 takes in the Kermadec Islands, but is not fished.

lift plummeted to levels almost as low as they had been before the intervention. Subsequent stock assessments[3] (together with the record of puerulus settlement[4]) showed that the high CPUE in the mid-1990s was brought about mainly by a large pulse of recruitment that had taken place along almost the entire east coast of central New Zealand a few years earlier, during a strong El Niño period. Once this peak had passed through the fishery, catch rates fell precipitously.

On the other hand, the Namibian fishery, which had produced as much as 9000 tonnes each year from the early 1920s but had then declined so much that that the minimum legal size was abolished in 1968–69 in an attempt to uphold landings, had remained at low levels.[5] In contrast to the assertion that there had been no

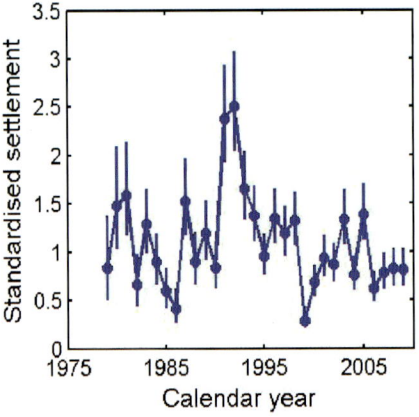

A large pulse of *Jasus edwardsii* puerulus settlement took place along the east coast of central New Zealand in the early 1990s during a prolonged El Niño, followed soon after by among the lowest settlements ever seen there. The error bars show the 5th and 95th percentiles.

management action, a suite of hard-hitting conservation measures had, in fact, been introduced from 1970 on.[6] But they failed to reinvigorate the fishery. This was largely because the Namibian fishery is greatly affected by coastal upwelling, which often leads to oxygen depletion and, in turn, to the demise of much of the inshore marine life, including spiny lobsters, as we saw at the end of Chapter 1. The tightly managed fishery for the same species just across the border in South Africa, part of the same lobster stock, underwent a similar dramatic decline.[6]

Researchers and managers are now aware that the abundance of spiny lobsters—and indeed many marine stocks—is not solely the result of fishing pressure and fishery regulation. Management intervention certainly can affect the fortunes of spiny lobster fisheries, but the Gisborne and Namibian fisheries show how relatively powerless such intercession can be in the face of natural processes and environmental variability. A major source of environmental variability is El Niño and the Southern Oscillation (ENSO). The discovery by Alan Pearce and Bruce Phillips of strong correlations between levels of settlement of the western rock lobster *Panulirus cygnus* along the coast of Western Australia and ENSO events was pivotal in enhancing our understanding of spiny lobster recruitment.[7] Not only did it add spiny lobsters to the ranks of those species whose abundance is heavily influenced by changes in the ocean climate, but it also provided one of the most influential insights of our generation into the drivers of change in spiny lobster fisheries. Bruce Phillips and colleagues were later to conclude that 'the level of recruitment to the western rock lobster fishery in any year is related to levels of puerulus settlement, which in turn are linked with oceanographic processes operating towards the end of the oceanic migration of the larval and puerulus stages.'[8] Settlement levels are highly correlated with commercial catches 3–4 years later, which means that they can be used to predict catches and to provide insight into the causes of any downturns in the fishery. Naturally, everyone in the industry relishes this powerful tool.

The ENSO phenomenon works something like this. As explained by Allan Clarke of Florida State University,[9] for most of the year Peruvian coastal waters are cold because of a north-flowing surface current. But beginning around Christmas, the current flows south and the water warms. This warm flow has become known as the current of El Niño (Spanish for 'the child Jesus') because of its timing. Occasionally the southward flow and warm waters last longer than a few months, well into the following year. Particularly strong El Niño currents bring rain, so much that pasture covers the coastal desert.

We now know that the coastal sea surface temperature changes associated with the El Niño current occur not only near the Peruvian coast, but also extend along the Equator one quarter of the way around Earth. These temperature changes are caused by changes in the equatorial Pacific westerly winds, which are in turn related to a huge seesaw in surface atmospheric pressure that rocks slowly and irregularly every 2–5 years. One end of the seesaw is the tropical west Pacific, Australia, India and the east Indian Ocean. The other is the east Pacific. When the east Pacific has anomalously low surface pressure, the other end has anomalously high surface pressure, and vice versa—the Southern Oscillation.[9] The Southern Oscillation Index is a measure of this anomaly.

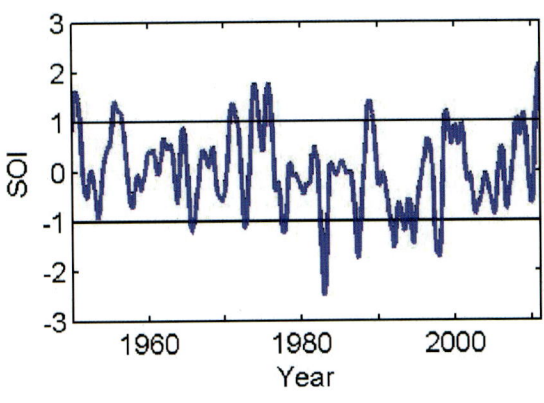

Southern Oscillation Index for the past 60 years. El Niño conditions exist when the index is negative.

So, during an El Niño, the surface waters of the east equatorial Pacific are warmer than usual and the equatorial trade winds weaker. The reverse is the case during La Niña (Spanish for 'the girl') conditions. Weakened equatorial trade winds result in fewer nutrients being available for the phytoplankton in the east Pacific, and consequently a marked decrease in marine plant and animal life there.[9]

But ENSO effects on the abundance of marine creatures in general—and spiny lobster recruitment in particular—are to be found well beyond equatorial seas, as we have already seen in the case of the western rock lobster. There is evidence that ENSO also affects settlement strength and/or catch size of the Hawaiian spiny lobster,[10] the red rock lobster along the east coast of New Zealand,[11] the California spiny lobster,[12] and the Japanese spiny lobster.[13] Other such relationships will almost certainly be detected as research continues. Invariably the mechanism is obscure. It may be that the increased frequency and strength of southwest winds south of the Equator during El Niños alter the patterns of water flow—and therefore larval distributions. Alternatively, they may bring about changes in the abundance and distribution of the phyllosomas' food, or their predators. The exciting point about all this is that, for spiny lobster stocks with a strong relationship between ENSO and settlement, knowing the state of the Southern Oscillation Index may be all that is required in order to predict the levels of recruitment to the fishery.

An interesting property of the Southern Oscillation Index is that it can remain, on average, high positive, neutral or high negative for a decade or more.[14] For example, the 1980s and 1990s were high negative (El Niño) years, the 1970s high positive (La Niña) years, and the 1950s and 1960s, together with the 2000s (until the strong La Niña beginning in 2010), were fairly neutral years. Herein may lie the main cause of past long-term variations in fishery performance; that is, the changes in lobster abundance brought about by variations in the ocean climate overwhelm those attributable to anything else.

φφφ

Sustainability is the buzzword of the primary sector. Not surprisingly, then, sustainable utilisation is the implicit or stated aim of the managers of essentially all spiny lobster fisheries. To this end, virtually all fisheries are governed by management measures, broadly divisible into input controls—those that direct the way fishers go about their fishing; and output controls—those that manage removals from the fished population. Almost all spiny lobster fisheries have input controls, but far fewer employ output controls.

Input controls that are widely applied include prohibitions on harvesting lobsters less than the minimum legal size (MLS) and females that are berried (carrying external eggs). Others applying to packhorse in New Zealand include making it illegal to commercially dive for or spear them, or to harvest soft-shelled lobsters; requiring pots to have gaps or mesh large enough to allow small lobsters to escape; and requiring the lobsters to be landed ashore live. Many fisheries also have both vessel and pot limits, and closed seasons—but the packhorse fishery in New Zealand is not among them.

The MLS first imposed in most fisheries was typically based on the smallest lobster size that markets would accept, but in some instances these were later modified to maximise yield. In most, but not all stocks, the MLS also allows most females to breed at least once before becoming legal to harvest. Packhorse fisheries in both New Zealand and Australia are exceptions. The MLS in New Zealand (216 millimetres tail length, which for females is equivalent to a carapace length of 155 millimetres) turns out to be the average size at which setae on the pleopods first appear, as discussed in Chapter 6. But the first clutch of eggs is not produced until the females are significantly larger (around 160 millimetres carapace length), and about 2 years older. In Australia, the MLS is a miniscule 104 millimetres carapace length, an awkward hangover with a long history. A recent management paper reports:

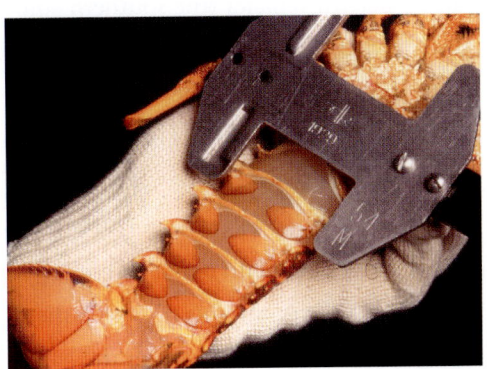

Minimum legal size for the red rock lobster is a tail-width measurement.

> There seems to be a degree of comfort with the current minimum size arrangements. This is most likely due to the fact that the benefits of an increase in LML [legal minimum length] to the stock are likely to be relatively much less than the economic costs that would be incurred by not being able to target the currently available smaller, more valuable lobsters.[15]

Because, potentially, not a single lobster in New South Wales reaches breeding size before being legal to harvest, there is great sensitivity about ensuring that sufficient breeders remain on the grounds. For this reason, a maximum legal size of 200 millimetres carapace length was introduced in 1994 to protect the large breeders. The size was reduced further in 2004, to 180 millimetres carapace length, for even greater traction. And it seems to be working. The spawning stock has increased to 0.26 of the unfished level and is expected to soon reach the target 0.30.[15]

But the more notable case of the MLS being smaller than the size at onset of breeding is the huge fishery for the western rock lobster in Western Australia. It used to be that in all areas except the Abrolhos Islands in the north, females reached MLS before beginning to breed.[16] Although this has changed to some extent, with more breeding now taking place along the mainland coast, things are still thought to be on a knife edge. Annual surveys of the grounds, conducted independently of the commercial fishery, provide managers with the critical information they need on what is happening to the extent of their breeding stock. Also, to help maintain egg production and to ensure that the breeding stock is not composed of just one or two year classes, a maximum legal size protects large females.

The prohibition on the taking of females with external eggs is an interesting one. It's hard to think of a regulated spiny lobster fishery where it is legal to harvest berried females (the southern spiny lobster *Palinurus gilchristi* off South Africa is one). Prohibiting their harvest tends to be a measure popular to both fishers and the public in general. But is it an indispensable management tool? What is the difference between a female that on one day had her eggs inside and the next day bore them externally? One day she was legal to harvest and the next day she wasn't. (In fact, it's not uncommon for females being held live before marketing to come into berry—so becoming illegal to possess—while in holding tanks. The way around this is to keep the males away from the females. Eggs that are not fertilised fail to attach to the hairs on the swimmerets under the tail and fall innocuously to the floor of the tank.) But of course, once protected from harvest, the legal-sized berried female has the opportunity to have that one extra clutch of larvae—which may or (probably) may not make a jot of difference to subsequent levels of settlement and, therefore, to the health of the stock.

Simplistic? Probably, because the real value in protecting berried females from harvest may lie more in the reduced overall fishing pressure on the lobsters that it brings about on the side. The level of effort reduction depends on how often the particular lobster species lays eggs, and on how long the eggs remain attached before hatching. For packhorse and most, if not all, subtropical and temperate species, it is once per year, over a few to several months.

There is one further difficulty with the ban on berried females. Hatching is never a completely clear-cut and clean process, and there are usually many infertile eggs remaining among the egg capsules. Even the most law-abiding of fishers can be caught out, the compliance officer using her hand lens to point out the one or two adherent eggs. To market these lobsters without prosecution, fishers may give the lobster a bit of a hand to get rid of her irritating remnants, by 'scrubbing'. (Some try doing this earlier in the egg-bearing season too, when the females are carrying their full clutch. But it's really difficult to get rid of all the eggs without leaving tell-tale evidence of iniquity.)

Closely related, of course, is a maximum legal size. 'You don't slaughter your breeding ewes or cows, do you?' is a common cry when lobster management is being debated. But only a few spiny lobster fisheries are managed with a maximum legal size, among the best known being the eastern and western rock lobsters in Australia. The rationale for a maximum legal size is that the larger females produce the most eggs so they may be viewed as being the most productive breeders.

But several important issues need be taken into account before introducing this measure. First, fishing pressure must be modest enough to allow sufficient juveniles to reach the larger sizes. Calls for a maximum legal size frequently bring pleas for a smaller MLS, with access to the smaller (and often immature) lobsters being seen as compensation for reduced access to the large lobsters. Next, the argument for a maximum legal size is most compelling where it has been established that greater numbers of eggs lead directly to greater numbers of new recruits. Such a correlation is by no means clear for any spiny lobster fishery—except, of course, that no breeding stock results in no recruitment. And the maximum legal size

The red rock lobster is the basis of lucrative fisheries in New Zealand and southern Australia, in the early days yielding enormous catches.

should apply to both sexes, because it is increasingly evident that large females require large males to fertilise all their eggs. Then there is evidence that larger lobsters restrict access to pots by smaller ones and that they can be cannibalistic. And finally, there is, not surprisingly, suggestion of reduced fertility among larger, older females—lots of eggs, but not necessarily fertile ones.[17]

Output controls constrain harvests in a quarter of the world's top 20 spiny lobster fisheries. The usual measure is a limit to the commercial harvest ('quota' or 'Total Allowable Catch') for the fishery as a whole. Sometimes the Total Allowable Catch (TAC) is apportioned to the different fishing groups (commercial, recreational, customary), leading to a 'Total Allowable Commercial Catch' (TACC). In a very few high-value spiny lobster fisheries, property rights to the resource (mainly transferable quotas) have been allocated to individual fishers. This is the way that the packhorse and red rock lobster in New Zealand are managed today. The Total Allowable Commercial Catch for packhorse is 40.3 tonnes, with just one fishery area, PHC 1, taking in all of the country.

φφφ

Sustainable utilisation requires information about the size of the stocks and how they have varied over time. But assessing the abundance of marine stocks—by sex and maturity, as is usually necessary—is a very much more complicated affair than counting the numbers of lambs, ewes and rams in a paddock. And sampling lobster stocks, probably with pots, is fraught with the problem of the catchability of individual lobsters varying for all manner of reasons—to say nothing about the costs of doing so on any meaningful scale if it is to be independent of the commercial fishing fleet. A considerable range of assessment techniques are used for spiny lobsters;[18] the more commonly used techniques—and a couple of the less-used—are discussed below.

Fishery Catch per Unit Effort
Such data as kilograms of lobsters per pot lift are often the only numbers available from the commercial fleet. The assumption—and it is only that—is that catch per unit effort (CPUE) is proportional to the abundance of lobsters on the grounds. Changes in CPUE can, indeed, be a useful indicator of trends in lobster numbers and are often used as a crude index of abundance in stock production models.[19] But there is the problem of comparing a unit of effort now to a unit of effort in the past, because improved gear and fishing technologies have increased the effectiveness of a pot lift. Also, spiny lobster catchability varies enormously according to such things as bait type, the lobster's stage in its moult cycle, environmental influences such as the phase of the moon and the season, pot-shyness by certain individuals, large lobsters excluding small ones, pot saturation, escapes from the pot, and so on. These can be dealt with to some extent by the mathematical process of standardisation (which tries to ensure you are comparing apples with apples), but often there is little or no detailed information on just how much each factor affects catchability.

And that's not the end to it. Typically, a fishery consists of many smaller groups of lobsters. These can be sequentially fished down with little reduction in overall CPUE over time—so long as new areas containing lobsters are found. High catch rates continue until the last accumulations are harvested, when CPUE suddenly declines dramatically. Managers must therefore follow CPUE separately across all parts of the fishery, and not rely on an overall CPUE.

For these reasons, CPUE data can be poor indicators of actual abundance, particularly when they have come from only the commercial fleet.

Fishery-Independent Catch per Unit Effort
More useful CPUE data come from dedicated research surveys broadly applied across the fishery, using standard gear, statistically optimised, and carried out at times of the year that avoid seasonal effects. Their main advantage is that they

Perhaps this is one of the few packhorse pots that Bill Hopkins on FV *Toiler* brought up close to empty.

are free from the location bias that tends to characterise commercial CPUE. They are, however, expensive—and usually there are still all those other problems mentioned above of using pot catches at all to estimate abundance. Accordingly, there are few programmes to survey spiny lobster abundance on the grounds. In the South African fishery for the Cape rock lobster, an annual pot survey provides indices of population structure and relative abundance.[6] Annual pot surveys of egg production for the western rock lobster, in Western Australia, are conducted over a 10-day period before the start of the commercial fishing season,[16] and there are also annual surveys of the size of the spawning stock of *Sagmariasus verreauxi* in New South Wales.[15]

Size-Frequency Distributions

Estimates of the total mortality rate (Z)—the sum of the fishing mortality (F) and natural mortality (M)—can be derived from the sizes of individuals in a fishery. (Refer to the previous chapter for an example, from the Far North packhorse fishery, of how fishing has altered the size structure of the catch.) These are obtained through sampling commercial catches (catch samples) or commercial landings (market samples).[19] Catch samples are more useful because lobsters smaller than the MLS, and others such as egg-bearing females, are included. It is usually F—essentially fishing pressure—that is of interest. But F cannot readily be separated out from M, and M is very difficult to determine empirically. In an unfished stock (where $F=0$), total mortality is equal to natural mortality ($Z=M$). M is usually assumed to be low, near 0.1 (which is roughly 10 per cent) and therefore only a small part of total mortality in heavily exploited spiny lobster fisheries. Knowing Z therefore allows managers to adjust fishing pressure to move stocks towards a preferred level of F.

But such things are seldom straightforward. Many factors such as gear selectivity, migration and catchability affect the size composition of catches. Seldom is it known how well the size and sex characteristics of the catches compare with those of the lobsters on the fishing ground. Nevertheless, as with CPUE, changes in the estimated value of Z can be useful indicators of trends in stock size.[19]

Direct Assessment

Dive surveys allow direct, fishery-independent estimates of stock abundance, but they are limited to shallow waters and are the most labour-intensive of all methods considered here. The best known is in Torres Strait, where the abundance of the ornate spiny lobster *Panulirus ornatus* has been estimated by divers for many years along transects sampling an area covering 25 000 square kilometres.[12] These annual surveys of relative stock abundance, together with catch sampling, are used to estimate the potential yield one year in advance.

Puerulus Settlement and Pre-Recruit Indices

Knowing how the numbers of particular early life-history stages—such as pueruli or pre-recruits (which are those lobsters coming up to MLS)—vary over time provides insight into the relative strength of annual cohorts that will later become part of the fishery. They also help in understanding the health of stocks. Any systematic decline in their abundance that cannot otherwise be explained through such things as fluctuations in the ocean climate may be because there is recruitment overfishing—the breeding stock is being fished so heavily that it is unable to replenish itself.

Settlement is usually gauged from specially designed collectors, for it is seldom possible to follow absolute puerulus numbers in nature. And there is a wondrous assortment of contraptions used around the world to do this.[20] They are of two main types: collectors that provide crevices for the settling pueruli, and those that mimic the structural complexity of seaweed. What characterises them all is their relative simplicity and ease on the pocket, being variously constructed of timber, air conditioning 'hogs-hair' filters, rope, and artificial seaweed.

The first collector developed was the Witham collector, used to catch pueruli of the Caribbean spiny lobster *Panulirus argus* in Florida. It closely resembled its successor, the hogs-hair collector illustrated opposite and now widely employed in Florida. The collector first deployed specifically to monitor levels of puerulus settlement over time was the artificial seaweed collector developed in the late 1960s for the western rock lobster off Western Australia. This design is now also used in Cuba for the Caribbean spiny lobster, and a variant is used to catch the pueruli of *Sagmariasus verreauxi* in New South Wales. The crevice collector is widely employed throughout much of New Zealand and southern Australia to follow settlement levels of *Jasus edwardsii*.

A strong relationship has been demonstrated between the levels of settlement of pueruli on collectors and the catches or catch rates from the fishery some number of years later for the western rock lobster,[21] and *J. edwardsii* in New Zealand[4] and Tasmania[22]. The relative strength of year classes, and therefore recruitment to the fishery, can be predicted. Correlations such as this tend to be most obvious for species and stocks where the lobsters reach MLS only a few years after settlement. The fewer the years, the less the 'smudging' of year classes brought about by variability in the growth of individuals. But even where there are many years between settlement and recruitment to the fishery, the settlement signal can still be powerfully persuasive, particularly if there have been strong trends in settlement over time—and especially if there is recruitment overfishing.

Fishery predictions based on indices of puerulus settlement from collectors can be validated and fine-tuned by incorporating indices of abundance of undersized lobsters, including pre-recruits—as they do in Western Australia.[21] And pre-recruit numbers in their own right can be used to predict catches: in Cuba, the annual index of abundance of juvenile Caribbean spiny lobsters is derived from monthly sampling of artificial reefs in nursery areas. More widespread estimates of pre-recruit abundance, however, most often come from catch sampling and log book data from commercial potting. (Such data tend to be based on the catches of only a few fishers over limited areas, or small numbers of lobsters in log books, and may not be representative of catches of the full fleet for the larger region, let alone of the actual sizes of pre-recruit and recruited lobsters on the grounds.)

The benefits of more than 40 years of remarkably accurate predictions of recruitment to the fishery based on the settlement of pueruli on collectors have

Managing the fortunes of fisheries

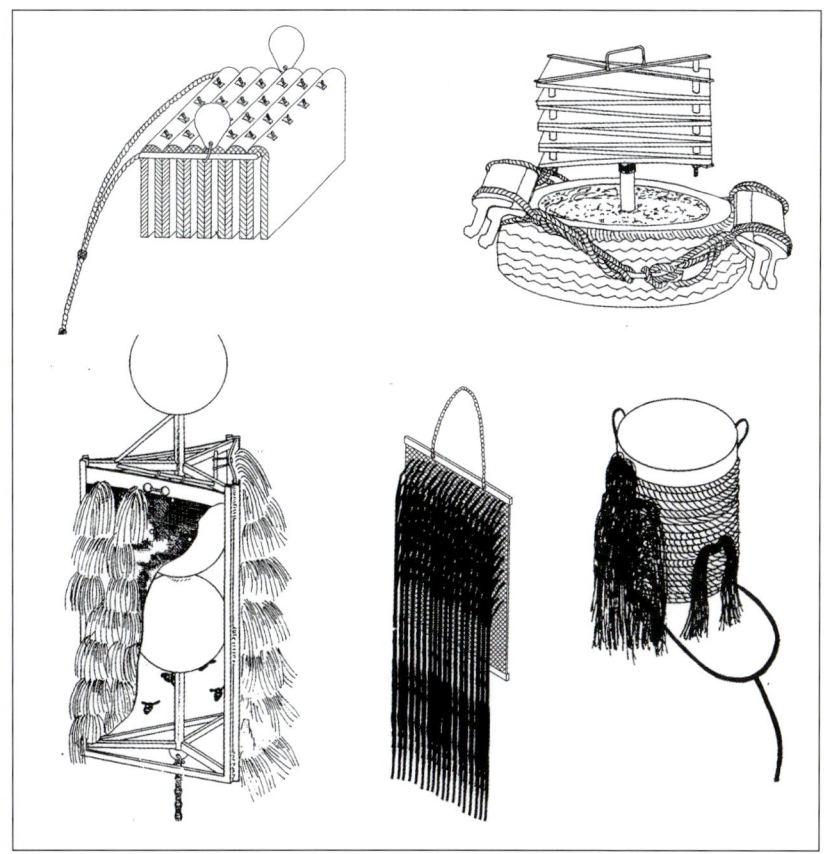

Examples of collectors used to estimate levels of puerulus settlement.[20] The hogs-hair collector for the Caribbean spiny lobster (upper left) and crevice collector for the red rock lobster (upper right) provide crevice-like conditions. (The hogs-hair collector is also fibrous, so in a sense imitates seaweed.) The lower three are artificial seaweed collectors for the western rock lobster (left), the Japanese spiny lobster (middle) and the Caribbean spiny lobster (right). The collector used for *Sagmariasus verreauxi* in New South Wales, Australia is a 2-sided version of the one at lower left.

The crevice collector (left) is used to index levels of settlement of the red rock lobster in New Zealand and Australia. The pueruli settle and remain in the collector until they are about to moult into the second juvenile instar (right).

been well demonstrated for the western rock lobster, the world's largest single-jurisdiction spiny lobster fishery. When the indices of puerulus settlement are combined with those of the juveniles, the joint index almost perfectly matches the fishery catches that follow.[21] The forecasts allow timely financial planning and investment by fishers and processors, and proactive rather than reactive fisheries management. All in all, a very good place in which to be.

Which is why perhaps no others in the spiny lobster world wait with more bated breath about what will happen to their fishery than the Western Australians. Settlement since the 2006–07 season has been among the lowest ever seen. The seasonal catch of each collector, which had typically been 50–150 pueruli, plummeted to just 10–30 pueruli and so far there has been no sure sign of improvement. There has never been such an extended period of poor settlement since records began. Particularly worrying is that most of this has taken place under the reign of a fairly neutral Southern Oscillation Index, rather than the El Niño conditions that typically lead to low settlement along this coast. Has there been recruitment overfishing? You can follow the drama month by month on http://www.fish.wa.gov.au/docs/pub/PuerulusSettlement/index.php?0405.

φφφ

The models most widely used to assess the status of fisheries these days are length-based because of the problems with aging spiny lobsters. They estimate the size of the current stock relative to such reference points as B_{MSY} (the average biomass—or weight—of lobsters that results from taking an average catch of maximum sustainable yield under various harvest strategies), B_0 (or virgin biomass, the average biomass of the lobster stock in the years before fishing started and therefore the stock's theoretical carrying capacity), or the biomass of lobsters at or above MLS in some particular previous period thought to represent 'a good place to be'. These models are technically very complex. Typically they use data on lobster sizes, growth rates, and catch rates. Management procedures, often with accompanying decision rules agreed to in advance by managers and stakeholders, may be used to translate the computed stock size or status into decisions that dictate whether the level of allowable harvest in the forthcoming season(s) is to be held as it is, or increased or decreased—and, if so, by how much.

The white digestive gland of three of these newly settled red rock lobster pueruli can be clearly seen. The puerulus on the right probably settled the night before it was collected, the others a few days earlier.

One of the most recent models is the multi-stock, length-based Bayesian assessment model developed by Vivian Haist and colleagues for the red rock lobster in New Zealand.[23] Bayesian assessments allow incorporation of prior beliefs, in this case that there is, for example, significant migration between stocks. State-of-the-art models of this type had been applied for some time to spiny lobster stocks in countries as far-flung as South Africa, the United States and Australia, but this is perhaps the most advanced iteration. It allows more than one stock to be assessed at the one time, with some parameters in common and others specific to each stock—potentially giving a better-informed result than a single-stock assessment. Recruitment to the model is also stock-specific, and the model allows variables such as migration between stocks to be estimated. It can also cope with growth rates that vary with the density of lobsters in different management areas.

This model was used in the joint 2006 assessment of two lobster stocks in southern New Zealand, CRA 7 (Otago) and CRA 8 (Southern). Each stock was also assessed separately. The model was driven by catch data (commercial, illegal, recreational and customary), standardised commercial CPUE, length-frequency data from the commercial fishery, and recent growth data. Changes in management, such as alterations in the minimum legal size, were incorporated. The model was run from 1976, which is as far back as reliable CPUE data extended, through to 2005. The assessment concluded that both stocks were healthy, the current and projected biomasses being well above the target and reference levels. The recruitment deviations pointed to a large increase in recruitment a few years into the new millennium. Projections were that the high catch rates would continue, with only a 25 per cent chance of decrease in biomass within the near future.[23] Accordingly, by the beginning of the 2009–10 season, the TACC for CRA 7 had been increased by 50 per cent, from 124 to 189 tonnes.

φφφ

Despite the availability of sophisticated assessment models like this, many spiny lobster stocks around the globe appear to be in trouble. Published commentaries suggest that close to half of the world's most important 20 or so fisheries have all the hallmarks of being overfished, and are no longer producing to their maximum sustainable yield.[24] For each there was the normal fish-down period of a newly discovered pristine stock, followed by a lengthy period of relatively high sustained landings. Then came disturbing declines in catches and catch rates, with the fishery being unable to produce anything like what it had been yielding. For example, the Brazilian combined fishery for the Caribbean spiny lobster and the smoothtail spiny lobster *Panulirus laevicauda*, once among the top spiny lobster fisheries in the world with annual landings up to 11 000 tonnes, now produces not much more than half this each year.[25] This may be an example of that very worst kind of overfishing—recruitment overfishing. But it could also have something to do with the less perilous, and probably more prevalent, growth overfishing. This is where the lobsters are harvested too small and too young, leading to yields lower than those possible had they been left in the water to reach maturity (which is when growth rates naturally slump). The obvious way to increase production of spiny lobsters is to rehabilitate stocks by managing extractions. Recruitment overfishing may be rectified by an immediate embargo on fishing, and growth overfishing by increasing the MLS and/or reducing fishing pressure. If the diminished performance is to do with environmental influences such as changes in the ocean climate, and management is otherwise sound, then patience may be all that is required.

φφφ

This chapter started out by demonstrating how remarkably powerful natural processes can be in determining the fortunes of spiny lobster fisheries. An example was the CRA 3 (Gisborne) stock of the New Zealand red rock lobster, one which had been performing poorly but then suddenly improved markedly. The verdict at the time was that the management package imposed in 1993 had worked brilliantly, as reflected in increased CPUE and lobster size'.[2,26] However, later assessments, together with the fishery-independent puerulus settlement record shown early in this chapter, indicated that the change for the better was much more to do with the large pulse of puerulus recruitment that had taken place along much of the east coast of northern and central New Zealand during a prolonged El Niño than the introduction of any inspired management measures. Further, projections (shown on the next page) of the size of the recruited biomass based on the settlement levels were far more informative than the recruitment projections from the assessment model regarding the calamitous decline in catch rates that was to follow.[27]

This chapter ends with a remarkably similar affair, one in an entirely different New Zealand red rock lobster stock. Yet another large recruitment event had taken place, this time in CRA 7—the east coast South Island fishery that was the focus of the sophisticated assessment model just discussed. This fishery was performing poorly, and its participants were probably disinclined to make it too widely known that their CPUE was by far the lowest in the country.

Against the background of a long period of low puerulus catches on local collectors and low catch rates in the commercial fishery, the heaviest settlement ever seen in CRA 7 took place. Starting in 2000, it reached several times its previous long-term average in 2003.[4] This pulse of settlement was recorded not just at Moeraki in Otago, but also, with uncanny precision—and entirely consistently with the oceanography—far to the south, at Halfmoon Bay on the east coast of Stewart Island. As the cohort grew, it was progressively noticed by fishers in their increased catches of pre-recruits, and later, of course, in their escalating fishery CPUE.

But settlement crashed after 2000–03, and puerulus numbers fell back to their much lower long-term average. Very little settlement was registered on the collectors in the years that followed. And, as you would intuitively expect—especially in a heavily fished stock like this one—CRA 7 fishery performance continued to tally closely with changes in the annual settlement signal. Only now it was all steeply downhill.

The settlement index was very insightful about the CRA 3 fishery performance, and still is. There have been strong correlations between it and catch rates in

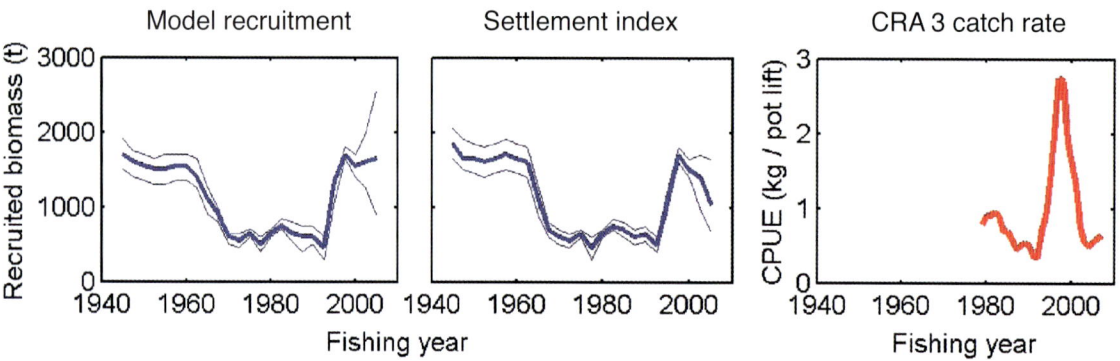

For the Gisborne (CRA 3) red rock lobster fishery, trajectories generated in 2001 of recruited spring-summer biomass based on model recruitments (left) and the Castlepoint puerulus settlement index (middle), the lighter lines indicating the 5th and 95th percentiles.[27] Castlepoint settlement was used as a proxy for Gisborne because the Gisborne settlement time series was too brief; both places are subject to essentially the same oceanographic regime and their indices are highly correlated. The settlement index was far more insightful of the collapse in the scaled, standardised CRA 3 catch rate that began in the late 1990s (right) — in spite of the introduction of a 'highly successful' management package.[2] (The total allowable commercial catch appeared to constrain landings in the 1993/94 to 2000/01, and 2008/09, fishing years, so the catch per pot lift those years may have been slightly different — either up or down — had that not been the case).[28] Not surprisingly, Gisborne catch rates continue to be highly correlated with local puerulus settlement levels.[4]

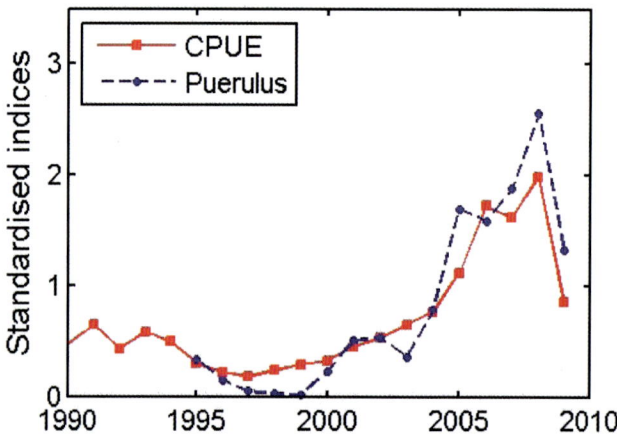

The Otago (CRA 7) red rock lobster fishery CPUE (catch per pot lift) plotted against the lagged local puerulus settlement index. To account for growth-rate variability between individuals, the settlement index was weighted so that for each year, 50 per cent of the index was for a lag of 5 years between settlement and the minimum legal size, and 25 per cent was for both 4-year and 6-year lags. (The total allowable commercial catch appeared to constrain landings in the 2002/03 and 2004/04 to 2007/08 fishing years, so the catch per pot lift those years may have been slightly different — either up or down — had that not been the case.) The 2006 CRA 7 stock assessment had predicted little chance of the biomass decreasing in the near future.[23] Settlement on collectors in Otago remained at or close to zero until 2008; low catch rates in the commercial fishery can therefore be expected until about 2013.

several other New Zealand fisheries, with lags that are entirely consistent with the field observations of growth rates.[4] (For those with correlations that were not statistically significant, plausible explanations are obvious: a short time-series of data for Wellington in CRA 4, and no clear trends in settlement for Chalky Inlet in CRA 8.) Most recently it accurately predicted changes in fishery performance in CRA 7—as powerfully and precisely as the Western Australians manage in their fishery. So, it is impossible to leave this topic without a note of negativity. As of June 2010, New Zealand's settlement indices have, essentially, never been used in the assessment of any spiny lobster stock.[*] 'Essentially' because the indices have been routinely dismissed by the assessors with reference to earlier reports that claimed the indices did not improve model fits. This is something like saying that if the settlement indices didn't precisely fit the model's view of recruitment then they were completely irrelevant. No ifs or buts. The state-of-the-art model is the only credible view of the world.

Does it much matter that the settlement indices have continued to be (essentially) ignored? Categorically, yes! Suppose you had been fishing CRA 7 in 2006. You and your family (and the multitude of local enterprises your fishing supported) could have reasonably concluded from the stock assessment that as long as everyone kept up the good work, high catch rates would most likely continue. There were lobsters on the reef—money in the bank—with a great chance of good numbers to come. So there was no good reason not, for example, to upgrade boat and gear then and there to make the most of the jackpot catch rates.

This in stark contrast to the clear conclusion of close to a 100 per cent expectation of decline in catch rates (probably a precipitous decline) if you had been shown the settlement index—the same settlement index that the stock assessors had at hand when they did their 2006 work.

The settlement indices seem to have been an anathema to the New Zealand rock lobster stock assessors, for reasons yet to be made clear. They seldom tried fitting their models to them as sensitivities, or using them as alternative forward projections of fishery biomass. Surely the assessors owed it to all those who depended on the fisheries to fairly use all available data—especially those

---

[*] The settlement indices are, I understand, now being used in stock assessments, but at the time of publication there were no publicly available outputs.

derived systematically and independently of the fisheries. The indices provide key insight into future fishery performance. At places with short intervals between settlement and recruitment, the predictions can be as precise year to year as they are in Western Australia; in others with longer lags, trends in catch rates are still predictable, though with less precision. And just imagine how more potent New Zealand's recruitment signals would now be had the effort put into discrediting them been devoted instead to making them even more powerful.

To end, cast your mind back to Western Australia where, after 4 years of inexplicably low settlement, the managers and industry have mobilised extraordinary measures. Like a nation at war, they have implemented previously unimaginable interventions. After long holding out against them, they have now applied an annual TAC (5500 tonnes—little more than half the long-term average catch), and quotas are on their way in. All this prompted largely by the unparalleled low settlements. Without any index of recruitment, everyone over there would have been caught completely off guard by their collapsing catch rates, their opportunity to do anything about it delayed by years.

CHAPTER 9

IMPACTS OF FISHING

There is no extant account. One can only imagine the seafloor of the Hole in the Wall in the northeast part of Mercury Bay, or the waters off Spirits Bay, one spring in, say, the early 1900s—well before any trawling or potting of packhorse. At the interstices between reef and sand, and out on the open seafloor, very large packhorse, sturdily intermoult and as happy as Eve, use just a frond of kelp for cover even during the day. Their size means they are prone to only the largest predators. Some of the lobsters are in waters so shallow that they are barely covered when the tide is out. These include the very small, dark green ones among the tribe—the result of settlement there the year before—which occupy weedy surfaces and crevices. The larger juveniles—many of them recent arrivals from the south—move between this verdant shelter and that provided by the groups of much older and larger packhorse out on the sand. Everyone forages most devotedly at night, the larger individuals working over the open seafloor for scallops on the surface and the just-embedded dog cockles, and others scouring the reef for gastropods and even other crustacea. The larger ones all moulted a month or so ago, and now there is to be mating.

All the while other life on the reef, the sand, and the interface between the two, goes about its business. What really strikes the eye are the fishes. There are the locals, including spotties and the more brightly coloured wrasses; porae with large fleshy lips, working over seafloor and rock surfaces in a manner reminiscent of coral-reef gnawers; the ubiquitous snapper, spotted in colours of the rainbow; the black-and-red vertically striped red moki near caverns; and the occasional large, sinisterly dark eagle ray on the sand just beyond the reef. Then there are the hordes just passing through. One minute a shadow darkens the entire seafloor as immense schools of shiny, small pelagic fish—koheru, jack mackerel—pass overhead, protected in part from predators by the 'confusion of the school'. Next there are dozens of giant kingfish—nemesis of the small pelagics because of their known penchant for them—with their dark upper side (difficult to see from above) and paler underside (not seen well from below), the flanks between filled by the longitudinal bands of green and brown that characterise this speedy, formidable predator.

What about now, a century later? At these and other northern places, the kingfish and snapper are smaller on average, and certainly less abundant. Most different—and, it must be said, most alarming, because of the magnitude and sheer extent of the change—is the monumental reduction in seaweed cover and the great proliferation of urchin barrens in the shallows, the genesis of which was discussed in Chapter 6. And for the packhorse themselves, a difference between then and now will also be obvious, particularly at locations towards the south of their range. This is where the small stocks of large individuals were always going to be far less resilient to fishing pressure than the much more substantial ones in the Far North. A dive back then at the Hole in the Wall, for instance, would have revealed a size composition something like a pair of camel's humps. There would have been a very well-established cohort of menacingly large adults (as we saw in the size-frequency distributions in Chapter 7), as well as a group of much smaller migrants. Nowadays, essentially all the packhorse are just the small ones passing through from the south.

About the only place these days where we can really get a hint of what life was once like on a lobster reef, before commercial fishing began, is within the perimeter of a reserve—where we may do little more than gaze. Worldwide, pressure grows to secure more marine protected areas (MPAs). Levels of protection vary—sometimes allowing, for example, certain types of fishing. But in this discussion the focus is on those where all fishing is banned. The purposes of such sanctuaries may be several, but within otherwise fished areas they are usually to rebuild and protect biodiversity, restore ecosystems to their pristine state, and to allow for scientific study.[1] It is also widely held that non-extractive MPAs increase the numbers and size of breeders of key fished species such as spiny lobsters, whose increased larval production helps to negate possible genetic consequences from fisheries selection in the harvested stock, and that they can even lead to enhanced larval recruitment.

The return of the balance of life on the seafloor toward something like what it was before fishing began is an obvious and reasonable expectation of MPAs—although there may never have been any such thing as a single enduring, steady-state climax community. This is because even in unfished waters there is constant change in community shape that reflects the varying fortunes of its denizens, some doing well one year and others the next. But certainly, allow the return of a full complement of lobsters and other keystone predators, such as snapper, to an area denuded of them and their presence will alter the balance forever. The reef will respond in a very noticeable way by reaching some other and most different equilibrium. Already such changes have been seen in the longer-established marine reserves in this country, most obviously in the revitalisation of the seaweed beds and diminished kina abundance.

The typical response of spiny lobsters when no longer subject to fishing is, not surprisingly, to grow in number and size.[2] This leads to increased numbers of breeding lobsters (spawning-stock biomass). Increased levels of lobster settlement and hence abundance in local—as well as surrounding—areas as a result of this is an attractive notion, but there is little if any empirical evidence for it.[3] For it to work this way there would need to be a strong, local stock–recruitment relationship, where increased numbers of eggs hatching lead to more-or-less proportional increases in subsequent lobster abundance on the same grounds. This is yet to be demonstrated for any spiny lobster, because of their very long and hazard-prone larval life, and the potential it brings for transport of larvae great distances away from where the breeders live, as well as the host of environmental factors that affect larval survival and, in turn, the abundance of new settlers. So, perhaps the most that can be claimed is that MPAs provide a potential buffer and antidote to heavy fishing pressure, and the larger they are, and the wider the network, the surer the medicine.

Because packhorse are so migratory, their rehabilitation within most MPAs will be very sluggish compared with less itinerant spiny lobsters. Gatherings of packhorse along the east coast of the North Island, such as at the Cavalli Islands, probably grew painstakingly, being made up of occasional lobsters that had decided enough was enough and not to migrate north any further. Indeed, after more than three decades free from fishing, no amassing of large packhorse has been reported in the Cape Rodney-Okakari Point (Leigh) Marine Reserve in the northeast of New Zealand. It is only in the Far North—the ultimate primary destination of the migrants—that an MPA could be expected to lead to significant increases in packhorse size and abundance within the short to medium term. And even then, to be truly effective the MPA would need to outwit the commercial pots and take in the entire range of the extensive seasonal inshore–offshore migrations. In the meantime, a network of MPAs does at least provide respite from fishing pressure for such migratory species as packhorse.

For most spiny lobster species, however, the upshot of protection from fishing—particularly where the fishing has been intensive—is dramatic. At four neighbouring reserves on the east coast of Northland, the density of red rock lobsters *Jasus edwardsii* increased 4 per cent each year in the shallows, and 10 per cent each year in deeper parts.[2] (At one of them—Leigh, established in the mid-1970s—lobsters are now perhaps twenty times more populous inside the reserve than outside it.)[1] The average size of the lobsters present increased by more than a millimetre in carapace length each year, and in biomass by 5 per cent each year in shallow areas and 11 per cent in deeper waters. Egg production increases mirrored those of biomass, or exceeded them because more and more of the population were large breeding females, with their larger clutches of eggs. It might not surprise us, then, to hear that Debbie Freeman of the Department of Conservation saw lobster densities within Te Tapuwae o Rongokako Marine Reserve, a little north of Gisborne, increasing at a staggering average rate of more than 30 per cent per year from its establishment in 1999.[4] Lobsters were shortly to be found wandering around in ankle-deep waters during the day, their antennae reaching up into space. And MPAs such as these can also bring direct benefit to commercial fishers. Where the population gets too large lobsters can be permanently forced away; the commercial fishers' expectation of this 'spillover' into surrounding fished areas is demonstrated by their precise delineation of the MPA boundary with their floats. The seasonal movements in and out of MPAs—such as large male red rock lobsters at Leigh moving onto open sand bottoms to feed on shellfish—are not spillover in the same sense, but unbothered by semantics, fishers just as keenly attend them.[5]

MPAs are now on the agenda of almost every country that boasts a coast. Nevertheless, their establishment is often bewilderingly controversial. We seem happy enough to set aside and protect areas on land to represent the local

Just a few years after Te Tapuwae o Rongokako Marine Reserve, on the east coast of the North Island of New Zealand (above, from foreground to the island at the top), was established in 1999, the red rock lobster became so abundant that large individuals could be found foraging on exposed intertidal reef platforms during the day (right).

biodiversity—such as a kauri forest in Northland—but there can be much more reticence and resistance when it comes to doing the same thing in the ocean. The main issue seems to be that the seas are often viewed as being communal property and that closing off areas to fishing is the denial of some right. (Individual quotas also often bring with them complicating issues of property rights and access.) Yet there is a more than ample case for completely protecting, at appropriate scales, representative marine ecosystems, together with those that are unique or otherwise special, in the same way as we do on land. And MPAs can themselves become popular and profitable crowd-pullers, especially when shallow and accessible. The Leigh Marine Reserve, made up of 500 hectares of mainly rocky shore, is visited by 300 000 people each year.[6] For it is in places such as this that it is possible to revere Earth's undersea creations and its creatures in more or less

The mainly rocky shore of the Cape Rodney-Okakari Point Marine Reserve near Leigh in northeast New Zealand is visited by thousands upon thousands each year.[6]

their natural state, however we may view their genesis. Even the most cynical may feel moved by spending time with Tane Mahuta's marine compatriots.

Many argue that MPAs should not be implemented for fishery management objectives, but University of British Columbia's Daniel Pauly takes an opposing view. He believes that no-take MPAs should be perceived 'not as scattered and small concessions to conservationist pressure, but as a legitimate and obvious [fisheries] management tool'.[7] Certainly they assist in countering the 'shifting baseline syndrome' referred to in Chapter 7, where only recent, and usually greatly diminished, levels of abundance of marine organisms tend to be accepted as normal and natural and you don't know how bad it has become because there's no one around who remembers how good it used to be.

φφφ

There can be less predictable consequences of fishing to our undersea neighbours, especially in terms of unintended catches. One can only imagine the comings and goings around a spiny lobster pot sitting secure and steady on the seafloor, before the sudden jerk that frightens all and takes the incumbents on their rapid journey to the surface. The ones not inside zoom in all directions. Those within can do little than cling—where they have the appendages to do so—or be forced against the mesh of the pot floor by the upward momentum.

I watched pots set for packhorse emerging from depths of 100 metres a few kilometres north of Hooper's Point, in the Far North, in April 2008. During hauling and before it is dropped to the deck, one pot was caught for a few seconds in the no-man's land at the air–sea interface: the roll of the vessel dunked the pot momentarily and then lifted it back into the air. At the moment the floor of the pot returned beneath the surface, the water was thrashed to foam by the last fish you would ever expect to see confined in a pot. Pelagic kingfish.

Some of the pots contained a dozen or so small (2-kilogram) kingfish *Seriola lalandi*, the most vigorous of detainees. There were similar numbers of the more ubiquitous, fleshy-lipped porae *Nemadactylus douglasi*. Altogether I recorded a dozen species of finfish (most commonly these two, as well as the carpet shark *Cephaloscyllium isabellum*), and numerous large hermit crabs of unknown provenance. Always the bait was long gone. Why were the fish there? The guess is that most if not all of them could have quite easily escaped from the pot, but were staying within to feed on whatever came along, and probably using the pot for cover.

Most other accounts of spiny lobster pot bycatch are generally consistent with this observation. For example, pots in Tasmania set for *Jasus edwardsii* have caught 30 fish species, 10 species of crustacean, and 10 other invertebrate species.[8] Escape gaps reduce the levels of bycatch significantly, and fortunately, most of the animals that reach the surface can be returned live—although they must of course get back down to where they came from without encountering a set of jaws on the way if an unchanged life-death ledger is to result.

There are other less expected (but perhaps more worrying) bycatches. At North Cape in late summer 1977, tangled in the line of one of our packhorse sampling pots, was a 1.6-metre-long leatherback

The surface of the water is turned white with the thrashing of kingfish in the pot.

The leatherback turtle, finally freed of its ensnarement, is prepared for winching back over the gunwale.

turtle *Dermochelys coriacea*, noisily gulping air as it surfaced. The largest and most oceanic of all turtles, these leatherbacks prey on the soft-bodied plankton that often accumulates around floating objects, and it is easy to see how their clumsy flipper movements could lead to a nasty tangle. Clearly the turtle had dragged the pot into deeper water in its bid to escape. Winched onto the deck, it was relieved of its rope windings, photographed and measured, and then with difficulty reunited with its more familiar milieu. Nat Davey once inadvertently snared a 1.5-metre-wide manta ray in a similar way in one of his packhorse pot lines off Spirits Bay.[9]

Other wholly inadvertent, but rare, bycatch victims of potting include species such as dolphins and whales, which can tangle in the ropes and drown. Unwelcome animals reported to occasionally enter pots in New Zealand waters, trying to rob either the bait or the lobsters themselves, include New Zealand fur seal pups and the endangered Chatham Island shag.[10] In Western Australia, pups of the rare Australian sea lion will also give it a go, most effectively being deterred by a rod projecting from the base of the pot up to the neck.[11]

φφφ

In terms of physical impact, there is hardly a more benign method of harvesting spiny lobsters than hand gathering. No illegal lobster need be touched. No other animal, or plant, need be interfered with. The greatest risk may lie in overharvesting—a matter for the managers. But hand gathering, although widespread in shallow tropical waters in particular, is rarely large-scale.

Potting is the method most extensively and intensively used to catch spiny lobsters and, fortunately, is also among the most selective and least destructive. Although they are far less perilous to the sea-floor ecology than tangle netting or trawling for lobsters (talked about in Chapter 7), pots are not, however, squeaky clean. Lobsters within pots—whether the pots are regularly hauled or not—are at greater danger from octopus predation than they would be if they were sequestered on the seafloor. Also, pots are often large and heavy—some weighing up to quarter of a tonne. The little hard data on their impacts (which mainly concern pots weighing less than 50 kilograms) indicate that pots cause the most damage to erect, bottom-dwelling creatures, such as corals and sponges, and to algae. They scrape, fragment, and dislodge these attached forms, particularly when dragged along reef edges and ledges during hauling.[12] And unfortunately corals and the like are usually slow-growing, long-lived, and crucial shapers of seabed communities because of the protection they bring to juveniles of other species and to

Department of Conservation's Debbie Freeman measures and records details about this red rock lobster in the Te Tapuwae o Rongokako Marine Reserve, before carefully returning it to where it came from.

other small life forms. The actual landing of the pots on the seafloor probably causes less severe damage to seafloor life than the subsequent snagging of lines and pots, and dragging in bad weather.[13] Overall, natural disturbances such as storms probably have a greater impact than pots on the ecology of shallow reefs, while in deeper waters the opposite is the case.[14]

Ghost fishing—pots lost by fishers (the buoys having become detached or submerged) and continuing to fish until they rust away, panels decay, or they become sufficiently buried in the bottom to cover their entrance ports—has generally not been seen as that much of a problem with single-chambered pots. Every fisher, exasperatingly, loses several each year, and those retrieved have invariably been empty.[15] Field and tanks studies with several species confirm that the lobsters are able to escape without mortality.

Or so the discussion has gone—until now. In a very recent report concerning the Florida fishery for the Caribbean spiny lobster *Panulirus argus*, where a staggering 50 000–100 000 pots are estimated to be lost each year, each pot is thought to effectively ghost-fish for up to a year or more.[12]

φφφ

An immediate and obvious impact of fishing is the culling of the larger individual spiny lobsters from the population, followed by great reductions in population density where fishing pressure is intense. With these changes in population structure come, theoretically, alterations in population dynamics. These include reduced size at maturity and reduced egg production for each individual, increased growth rates through less competition for resources, mating system anomalies, and potential changes in genetic structure.[3]

Another potential biological effect of fishing is greater susceptibility to, and incidence of, disease. Red rock lobsters in the intense commercial fishery outside the Te Tapuwae o Rongokako Marine Reserve were far more afflicted with tail fan necrosis that those within it.[16] Then there is reduced growth brought about through handling stress and injury.[17] Incidental, or indirect, mortality of sub-legal or otherwise non-takeable individuals during fishing is increasingly an issue as fisheries become more intense. For example, in poorly designed pots that sit flush on the deck, lobster limbs curled around the floor of the pot are crushed, and while lobsters are being removed from the pot there is often limb loss. Limbs can be replaced, but at a biological cost, and in the meantime the lobster is at greater risk from predators.[18] Sub-legal individuals may be caught many times during a fishing season, and each time there is the possibility of injury.

In most places, regulations require that the non-retainable part of the catch be returned to the sea without delay—for good reason. Spiny lobsters quickly weaken in air, especially when confronted by high temperatures and drying wind. And light-induced damage to the eyes' photoreceptors from long-term exposure to moderately bright light or brief exposure to very bright light—in extreme instances causing complete blindness—can severely disrupt normal functioning.[19]

Culled lobsters falling back through the water column to the seafloor are prone to predators. Certain fishes may even follow lobster vessels in anticipation of catching these unfortunates. And obviously, lobsters returned to a seafloor totally different to that normally occupied face a high risk of being eaten—especially those lobsters kept on deck too long and requiring several minutes on the seafloor to recover.

Ominous clouds gather on the horizon for many spiny lobster fisheries. The name of the game is not only resource management, but also, in these times of concern about climate change, carbon management. Carbon emissions from spiny lobster fishing are enormous compared with most other primary production, reflecting the energy-intense nature of potting and the high prices that make fishing viable at low catch rates. It has been estimated that for each kilogram produced, total carbon emissions from lamb production and from dairy production in New Zealand amount to 0.7 and 1.4 kilograms respectively. In contrast, the quota-managed part of the South Australian fishery for *Jasus edwardsii* emits 7.4 kilograms of carbon for every kilogram of lobster meat produced, and the part of the fishery managed only by inputs emits a whopping 20 kilograms—the carbon emissions in each case coming mostly from fuel use.[20] Such headline-grabbing levels are likely to bring about change in the way that spiny lobster fishers go about their business. For example, we may see further fisheries output-managed, more pots being fished with less frequent lifts, and so on.

The steadfast progress of climate change (very likely human-induced), which will take quite some effort to quell, let alone reverse, may mean that the distributions of spiny lobster species shift poleward as waters warm. This is because for all lobster species, some waters are just too hot. And the larval stages of spiny lobsters may be even more profoundly affected by rising temperatures. First, ocean currents may change as waters warm, bringing about altered patterns of larval dispersal, and even failure of some of the usual recruitment mechanisms. Second, acidification of the ocean, due to increased absorption of carbon dioxide in the sea as the air becomes more and more full of it, potentially affects all marine creatures with thin shells containing calcium—including phyllosomas.[21]

CHAPTER 10

GREEN FINGERS

Len Tong and I had waited ages for this. After the 10-hour flight from New Zealand, and then the at-times 250-kilometre-per-hour journey north from Tokyo by Shinkansen ('bullet train'), there had been hours sipping tea in austere rooms while Professor Jiro Kittaka was involved with other things. The starkness of the square hard-edged buildings of Kitasato University was relieved only here and there by brightly coloured signage. Concrete tanks, stained brown by time and use, on the flat area below the main buildings resembled rows of small swimming pools in need of attention. Yet it was down there that the breeding lobsters were housed. And in one corner of the field of tanks were a couple of plastic-covered greenhouses where history was being made.

Jiro Kittaka (left) displays Rick Webber's line drawing of the final-stage phyllosoma of the red rock lobster just given to him by the author (right) at the 7th International Conference and Workshop on Lobster Biology and Management in Hobart, Australia in February 2004. The presentation was to acknowledge Jiro's enormous contribution to our understanding of phyllosoma culture and behaviour.

Both of us had examined numerous pickled phyllosomas over the years—crumpled and dull, like used linen. But here in the culture containers in the greenhouse were live, looping, late-stage phyllosoma larvae, transparent and tumbling, hunting and holding prey. The occasional flashes of red were the chromatophores on their legs. What we were seeing that day in late 1988 were the earliest roots of the commercial production of spiny lobsters. Before us were larvae of both packhorse and red rock lobsters.

Aquaculture (the growing of a species on land or in the sea in a controlled or semi-controlled environment, or the ongrowing in captivity of a species taken from the wild at an earlier life-history phase) is one of the few methods of increasing spiny lobster production in the face of declining commercial catch rates. (Another method is enhancement, which is discussed later in this chapter.) As a pioneer in the world of spiny lobster aquaculture, no name stands above that of Jiro Kittaka. Working at Sanriku (sadly affected terribly by the 2011 tsunami), in Iwate Prefecture towards the north of the main island Honshu, and later at Nemuro in the north of Hokkaido, he was the first to grow any spiny lobster through its full larval development, to settlement. In the end he accomplished this for five species. During the 1980s and 1990s, Jiro grew first the Cape rock lobster *Jasus lalandii* from South Africa, then the red rock lobster and the packhorse from New Zealand and Australia, and the European spiny lobster *Palinurus elephas* from Ireland. He also cultured the local Japanese spiny lobster *Panulirus japonicus*.[1]

Jiro Kittaka and his team visited New Zealand several times to see the sorts of places where both packhorse and red rock lobster are found—information that would assist him in his quest to culture these lobsters back in Japan. Recently he supplied me with some reflections on quarter of a century of research into the very young spiny lobsters.[2] 'I particularly remember when we stayed in a small inn near Castlepoint one night. We were attacked by a group of flea,' he began.

> Many people were trying to raise the subtropical Japanese spiny lobster *Panulirus japonicus*—without success. Based on the northeast Pacific coast of Japan, well beyond the range of this species, I wondered whether cool-water spiny lobsters would be suitable for culture in my institution. In 1981 I organised a team to survey the ecological requirements of *Jasus* in the Southern Hemisphere, first visiting South Africa, then Australia and

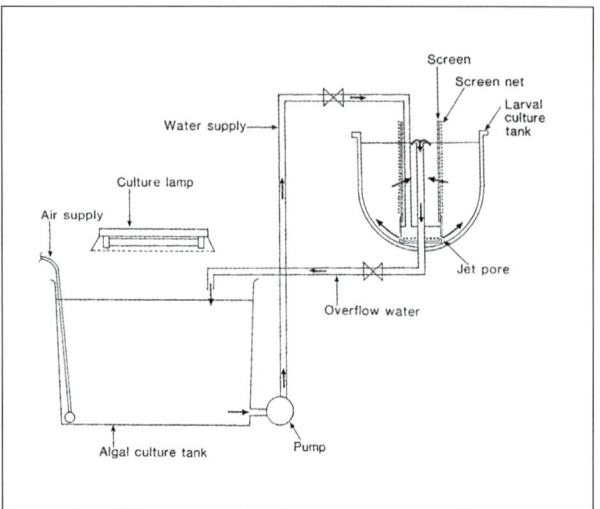

The set-up used by Jiro Kittaka to culture the phyllosomas of packhorse and four other spiny lobster species (left). The phyllosomas are in the tank, which appears green because of the microalgae that help control bacteria and maintain water quality. Larger tanks were used in later experiments but the system was essentially the same, as illustrated in the diagram above.[1]

New Zealand. Mature lobsters—both males and females—from all three countries were air-freighted to Japan.

A female *Jasus lalandii* moulted on 22 March 1986, then mated and spawned a day later. On 1 August 1986 the eggs started to hatch, and 16 000 of its phyllosomas were transferred to a culture tank. A single phyllosoma survived to metamorphose. This, the first puerulus ever cultured, arrived 306 days after hatching.

I ascribe my success in culturing phyllosomas to the use of small pieces of mussel flesh as the main food. We had noted how after we had added several drops of mussel extract into the culture the phyllosomas would come to the surface to gather every piece of mussel. Also, the microalgae inoculated in the culture tanks were important because they controlled water quality and pathogenic bacteria.

In 1987, using these techniques we cultured *Sagmariasus verreauxi* at Sanriku. These were very tough animals. We ended up with 168 pueruli, after 189–359 days. Later, 10 juveniles were shipped to Nemuro. All survived there, mated and carried fertilised eggs in 2002, a decade after they had hatched. Their larvae were in turn cultured to settlement, but before maturing they were lost in a laboratory accident. Nevertheless, two successive generations of larvae had been produced for the first time ever.

'*Jasus verreauxi* phyllosomas cultured' announced the *Lobster Newsletter* in May 2000.[3] New Zealand had become just the second country to culture any spiny lobster species to settlement, thanks to the work of National Institute of Water and Atmospheric Research (NIWA) scientists Len Tong, Graeme Moss and colleagues on both the packhorse and the red rock lobster. The packhorse phyllosomas had been fed live brine shrimp and grown at 21–24°C. Survival was high until mid-stage, when a virus caused the larval population to plummet. Nevertheless several pueruli were produced, an average 240 days after hatching.

Graeme Moss and his team went on to show how packhorse egg-hatching can be extended over several months by managing the temperature of the holding water. Cooler water retarded development of the embryos.[4] Further, with the egg extrusion date and holding temperature known, the time to hatching could be predicted. These discoveries were good news for aquaculturists because it meant they could exert quite a lot of control over when larvae were produced. Graeme's team also found that by manipulating temperature and food levels they could significantly improve survival and shorten intermoult periods.[5]

The approach most widely used to culture phyllosomas these days is to keep conditions as aseptic as possible, including the food provided. Batch culture in large tanks has been achieved. Water changes and filtration deal with dead larvae and uneaten food, and ultra-violet sterilisation and/or ozonation of the water helps to control bacteria and other unwanted microorganisms. And keeping the water moving is key to the larvae remaining in suspension. Once the phyllosomas fall to the bottom of the container—which they tend to do because of their aversion to light and because of gravity—they can damage their long thin appendages or become tangled in debris. Best results have been achieved under dim light in moderate currents that ensure that the larvae can swim freely and that they will frequently encounter drifting food items.

This packhorse puerulus could have become the largest spiny lobster ever.

Eleven spiny lobster species have now been cultured to settlement, all the recent ones being *Panulirus* species. And packhorse remains one of the most promising in our attempts to produce commercial quantities of small spiny lobsters for ongrowing. So far only it and the Japanese spiny lobster have been produced in anything but very small numbers, packhorse surviving the best. (There are rumours, however, of success with the ornate spiny lobster *Panulirus ornatus* in both private and joint ventures in Australia.) And the baton of leadership in packhorse culture has now decisively passed to Australia's University of Tasmania, and particularly to Arthur Ritar. Arthur describes the point reached by his team in their investigation into the production of packhorse phyllosomas and juveniles:[6]

> The packhorse has rightly been considered easier to culture than other temperate or subtropical species, possibly even tropicals, as they not only have a relatively short larval phase, but are also hardier. In 2006, we first cultured larvae from egg to metamorphosis, in 7–10 months at 21–23°C. Of the resulting 52 pueruli, those grown at 21°C reached 300 grams in 18 months. This growth rate, and their good survival, make packhorse an attractive proposition for commercial aquaculture. [Taste tests have demonstrated them to be at least as good as the highly regarded *Jasus edwardsii*.]
>
> Larvae are cultured in the hatchery under intensive conditions at densities far higher than in the wild, and at warmer temperatures. Unfortunately, the hatchery environment is conducive to the proliferation of disease organisms such as *Vibrio*, particularly when live brine shrimp, a known vector, are involved. Disease has hampered the culture of larvae, often causing catastrophic mortalities. Although antibiotics are highly effective in suppressing or eliminating potentially pathogenic bacteria, their use is regarded as unsustainable because resistant strains will probably emerge. Instead, we have developed a water treatment process relying on the ozonation of seawater. This produces an oxidative environment that minimises pathogens while ensuring larvae survive well and grow rapidly. Although the implications of chemical changes occurring with ozonation of seawater are poorly understood, the technique shows great promise for the culture of larval spiny lobsters, and probably other crustaceans as well. To date, we have used ozonation only in flow-through systems but in the longer term it will need to be applied to recirculated water as well.

Phyllosomas require soft, gelatinous or fleshy foods. However, the typical diet administered in the hatchery is vastly different to that in the wild. Under culture, the main food item has been live brine shrimp supplemented with pieces of fresh mussel gonad. The brine shrimp are grown from cysts to a sub-adult size of 7–10 millimetres before feeding to the larvae. Even small, newly-hatched phyllosoma are able to capture and eat these relatively large shrimp. Typically, the phyllosoma senses items of food with the tips of its appendages, then grasps and manipulates them to its mouth where they are torn, threshed and masticated to a slurry of fine particles before final passage and absorption in the digestive gland. But reliance on costly live feeds presently hampers mass hatchery production of seedstock for commercial aquaculture.

Metamorphosis of the phyllosoma larva to the postlarval puerulus requires a major expenditure of energy. Survival during this transition may be severely compromised by inadequate nutrition and husbandry during the long period of culture, resulting in the phyllosomas being unable to complete metamorphosis. This is an area in need of more research, a more thorough understanding of the physiology involved being required to ensure high survival through to the puerulus stage and beyond.

Clearly, there are many biological and technical challenges to address before the commercial aquaculture of packhorse using hatchery-reared seedstock becomes an economic reality. However, recent advances demonstrate that, through systematic research, it should be possible.

It seems pretty certain then that commercial culture of spiny lobsters from the egg to the juvenile will be achieved in the medium term. They are good candidates because they can mature and breed in captivity and grow well

Late-stage (Instar 14) packhorse phyllosomas (left), pueruli and first-instar juveniles (middle), and larger (300 grams) juveniles (right) cultured in Tasmania.

communally. Moving toward the goal of commercial-scale larval production has been expensive, with more than US$10 million having been spent in the past decade. But the end is in sight—a number of groups in several countries, including Australia, are now proceeding with plans for commercial hatcheries to raise seed spiny lobsters for aquaculture.[7]

The ultimate goal of phyllosoma culture is, of course, not to produce just pueruli and young juveniles, but rather, spiny lobsters large enough to sell. Lucrative markets exist in Asia and Europe for all sizes of whole live spiny lobster over about 250 grams in weight (cultured lobsters usually don't need to comply with the size rules of the wild fishery). There needs to be good-enough growth and survival of the small lobsters, right through until they reach saleable size. So, how did that growth rate of 300 grams in 18 months for cultivated packhorse compare with other cultured species? At 18 months, cultured western rock lobsters and California spiny lobsters *Panulirus interruptus*, both subtropical species like the packhorse, were 200 and 400 grams respectively.[8] Not that much in it. But at the same age, the tropical Caribbean spiny lobsters were 500 grams, and ornate spiny lobsters *Panulirus ornatus* even heavier. Clearly the tropical species have the edge in terms of growing the grams.

Irrespective of species, sourcing appropriate feeds, controlling disease, and the high costs of infrastructure and labour will be the greatest challenges to ongrowing juveniles at commercial scales. Spiny lobsters are rather inefficient feeders, with food conversion ratios (wet weight food consumed: increase in body weight) between 4:1 and 9:1 having been reported for natural foods.[8] Accordingly, food may comprise up to half the production costs. Because natural shell colour is important to many markets, its maintenance in cultured juveniles may require the addition to the diet of carotenoid pigments such as astaxanthin.

The most optimistic predictions for economic success come from low-labour-cost economies using tropical species such as the ornate spiny lobster and the scalloped spiny lobster *Panulirus homarus*. These lobsters have reasonably short larval durations and fast growth. They also have boldly coloured bodies that appeal to many customers. But two other species cannot be ignored as rivals. First, the packhorse, because of its moderately short phyllosoma phase, the demonstrated high survival of phyllosomas and pueruli, its hardiness at all phases of its life cycle, and its good juvenile growth rate. And then the European spiny lobster, which has a short larval life (because it hatches at an advanced stage) and fast-enough juvenile growth. Successful commercial culture of these slower-growing species may involve shore culture for the first few months, followed by less-expensive ongrowing to market size in sea-cages or in cages within channels of flowing water.

There is one further area of endeavour to note in the farming of spiny lobsters, with which University to Auckland's Andrew Jeffs has had first-hand experience.[7]

> It is possible to avoid the hatchery step for spiny lobster aquaculture by collecting early juvenile lobsters from the wild to provide the seed source for on-growing. A growing number of countries are adopting this approach as a way of increasing production from static lobster fisheries. Vietnam has been the most successful, producing around 2000–3000 tonnes of aquacultured lobsters a year starting from small juveniles of several tropical lobster species caught from the wild. These lobsters are taken in nets, collectors and by divers in areas where large numbers of early juveniles are known to dwell. The small lobsters are then transferred to farms where they are held in cages made of fine mesh which are submerged beneath floating rafts. The small lobsters are fed fresh chopped mussel, clam and fish meat. As the lobsters grow they are moved into larger cage nets slung beneath the moored rafts. Fast-growing species like the ornate spiny lobster can reach a market size of 1 kilogram in as few as 14 months and return more than US$60 each to the lobster farmer when sold live into key Asian seafood markets such as China.

In Vietnam, the pueruli of tropical spiny lobsters (centre) are caught in fine-meshed nets (left) and ongrown to market size in floating cages (right).

We leave the business of spiny lobster aquaculture on a somewhat ghoulish and controversial note: eyestalk ablation. Snipping off the lobsters' eyes at the base of their stalks can promote growth spectacularly, mainly through accelerating moult frequency (probably because the source of the moult-inhibiting hormone is a gland in the eyestalk). Eyestalk ablation increases the feeding rate and can also improve food conversion efficiency. Small juvenile scalloped spiny lobsters that have had their eyestalks removed can have weight gains three to seven times greater than those not ablated.[8] However, ablated lobsters often have lower survival than those intact, and there have been adverse behavioural and colour changes. They may also be more susceptible to poor water quality and reduced oxygen levels than whole animals.

Removing just one eyestalk can lead to better survival and yield than bilateral ablation or no ablation, even though growth may be slower. But the desirability of the practice, and the suitability for live sale of lobsters without eyes (or with artificial eyes), is yet to be seen.

φφφ

Enhancing marine resources is a hot topic, particularly for invertebrates.[9] It involves manipulating the physical and biological environment in order to increase harvests, or supplementing natural recruitment by introducing new stock. NIWA's Megan Oliver investigated two different approaches to enhancing populations of red rock lobsters: releasing early life-history phases into natural or modified habitats, and transferring lobsters from one habitat to another.[10]

> Juveniles released into the wild to augment the numbers about to recruit to the fishery may be raised in the hatchery, or caught in the wild at an earlier stage of development and then on-grown in captivity before release. On-growing juveniles in captivity can be useful for overcoming survival bottlenecks in the wild. For example, if juveniles are less vulnerable to predation once they reach a certain size, there is advantage to on-growing them to a larger size.
>
> Stock enhancement may also be achieved through the provision of artificial habitat designed to improve settlement of juveniles where habitat is limiting. Translocating juveniles or adults from areas of poor survival or growth to better areas also constitutes a form of stock enhancement.

Stock enhancement efforts have been directed, with mixed success and varying degrees of scientific rigour, at numerous fish and invertebrate species, including cod, flounder, seabass, scallops, abalone, clams, queen conch, sea cucumbers, sea urchins, shrimp, clawed lobsters and spiny lobsters. Packhorse's cousin, the highly valued *Jasus edwardsii*, has been the subject of several recent enhancement feasibility studies. Despite their gregarious nature and good survival in captivity, the protracted larval phase has so far made commercial-scale hatchery production of seed unfeasible. Therefore, the recently settled pueruli and young juveniles are trapped in the wild for on-growing in captivity.

Red rock lobster pueruli taken on crevice collectors at Gisborne were ongrown in captivity for enhancement experiments.

Maintaining animals in captivity for several months or more, however, potentially disrupts behavioural development and contributes to high mortality after release into the wild. But behavioural changes induced by on-growing in captivity do not appear to influence post-release behaviour or survival. Lobsters reverted to 'wild' behaviour almost immediately upon release, displaying good defensive behaviour and resuming nocturnal foraging patterns.[11] Reseeding should therefore lead to enhancement of wild populations where there has been careful selection of release sites and use of suitable release protocols.

Few commercial efforts have been made, however, to capture, on-grow and then release juvenile lobsters for the purposes of stock enhancement. Reasons for this include the immense effort required to trap tens of thousands of young juveniles, the cost of on-growing them in captivity, and issues of ownership and cost-recovery after release if lobsters move away from the release site. There has also been concern that diseases, both common and novel, may be more prevalent in captive-reared lobsters and could spread to wild populations.

Translocating adult lobsters as a supplementary management strategy in the Tasmanian *Jasus edwardsii* fishery has involved trapping and then moving wild individuals from one part of their range to another to establish, re-establish or augment the population.[12] Lobsters in deep-water regions of Tasmania are slow-growing, have narrow tails, spindly legs and are pale in colour. In contrast, lobsters from shallow-reef regions are faster-growing, larger, and an attractive deep-red colour, commanding a much higher export price. Furthermore shallow-water lobsters have substantially better survival rates during transport to overseas markets.

Pilot studies have indicated that translocating deep-water lobsters to shallow areas of reef does indeed promote the desired colour change and a higher mean growth rate. Results from subsequent large-scale translocations will clarify if this is a biologically and economically feasible means of stock enhancement through directly providing lobsters to the fishery and/or boosting biomass by increasing the number of spawning adults.

The most successful enhancement demonstrated (as yet only experimentally) for spiny lobsters anywhere is the provision of artificial shelters for settling pueruli and young juveniles of the Caribbean spiny lobster in Florida, to help protect them from predation.[13] It remains unclear if these shelters—simple, inexpensive structures that provide low-profile refuges for the lobsters—actually enhance survival of settlers or if they simply concentrate juveniles, as discussed with reference to pesqueros and casita cubanas in Chapter 7. For the moment though, the shelters do appear to help overcome the demographic bottleneck of too few refuges of the right form.

Perhaps the most extraordinary attempts at spiny lobster enhancement are in Japan, where the creation of leviathan artificial reefs composed of concrete blocks and stone beds, to enhance settlement and abundance of the Japanese spiny lobster, goes back to the 1930s.[14] The reefs are constructed in such a way as to provide refuges for settling pueruli and habitat for otherwise dispersed juveniles. One particular project to enhance coastal fisheries, begun in 1976, saw ten locations in eight prefectures established by 2000, each consisting of a fishing ground of about 100 hectares. It involved setting huge concrete blocks of different shapes atop stone beds, the blocks affecting water flow and promoting puerulus settlement among the growth on the surrounding stone beds. The beds accommodated these natural settlers, along with juveniles that had settled elsewhere. A slightly different approach was used in one of the prefectures (Shizuoka). Dozens of hollow 40-cubic metre concrete blocks, each with narrow slits over all sides, were placed on cobble tracks up to a kilometre long.[15] Here the concrete blocks were themselves artificial shelters for settling pueruli. Again, in spite of so much concrete and stone having been placed—possibly even enough to slightly upset the balance of rotation of the Earth—the effectiveness of these enhancement methods has not been conclusively demonstrated.

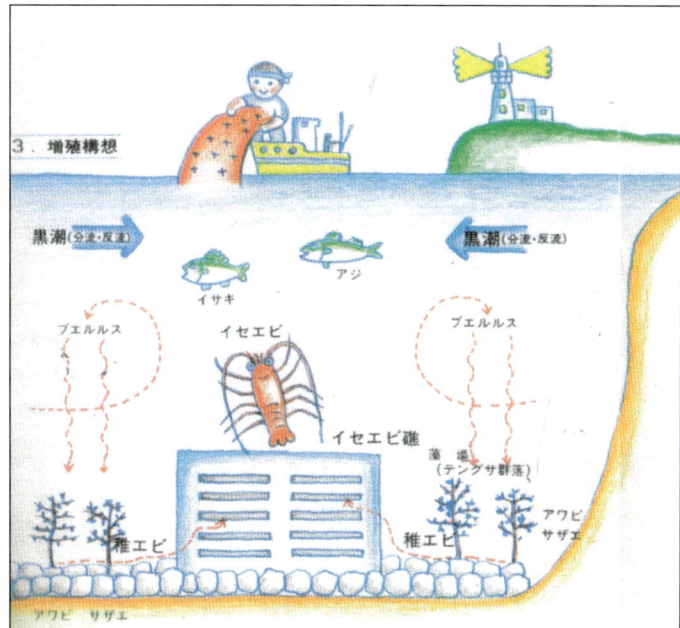
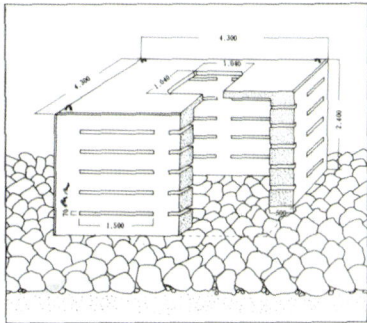
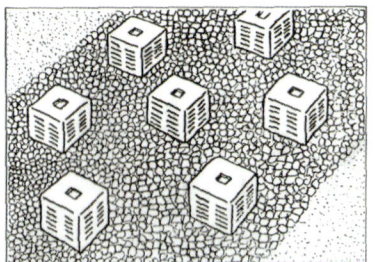

Artificial reefs to enhance the abundance of Japanese spiny lobsters in Shizuoka[15] on the east coast of Japan are on a massive scale but their effectiveness remains unclear. The red downward-facing arrows in the cartoon represent the settlement of pueruli onto red algae growing on the cobbles.

CHAPTER 11

THE FINAL CHAPTER

The story of packhorse and other spiny lobsters is, whether we like it or not, incomplete if we do not consider their ultimate destination. Of the few that survive the perils of being at sea for months as larvae and then settling in unfamiliar environments, most eventually find themselves being packed and marketed. And it is again their distinctive make-up and strange habits that make spiny lobsters marketable live virtually anywhere on the globe. Although they are creatures of the sea, they are remarkably capable of surviving without it.

Australian lobster biologist Bruce Phillips and I waited, saying little, in the early morning light outside the Tokyo Station Hotel. A block or two away, news crews with telecast trucks were maintaining a 24-hour vigil, ready to announce the death of the Emperor, who was ailing within the castle-like walls of the Imperial Palace. Meanwhile, large gaudy carp continued to swim back and forth in the moat that surrounds the stern stone walls, not the least bit interested in all the fuss.

We may have bemoaned, a little, the early hour, but Jiro Kittaka had had to start much earlier to get us to Tokyo's fish market in good time, before everything had disappeared. We still had a short train journey, and although the carriages were filling they were the merest hint of what they would become in a couple of hours. This is when, without dignity, you are shoved into the carriage by people employed to do just that; when almost certainly your party will be split because you simply won't all fit on the first train; where you may feel a rising wave of claustrophobia at the same time as you worry what you would do if someone had a heart attack or if a child was being crushed (which is absolutely nothing because you are so tightly bound); and where you must clearly understand that you need to barge your way to the door on the opposite side of the carriage—across the centre of this container of tightly squashed human-kind—in good time to be able to alight at your station. This imposed intimacy is not enjoyed by anyone, if the look on the sea of faces around you is anything to go by.

There's an astonishing array of live, chilled and frozen seafood from around the world to be seen at the Tokyo Metropolitan Central Wholesale Market (commonly known as the Tsukiji Market). Much of it is auctioned to middle-men who will on-sell to outlets as varied as the most elegant high-street restaurants and the most modest backstreet eateries. Giant frozen tuna from somewhere way out in the Pacific, appearing to smoke as they begin to thaw; live bivalve shellfish collected more locally and crammed into aerated tanks; chilled uni (sea urchin roe) in delicate wooden boxes. And both Bruce and I found tanks of live spiny lobsters from our respective countries.

It wasn't just the extraordinary price of the Japanese spiny lobsters *Panulirus japonicus* that astonished us, but also their very small size; some were barely larger than the palm of your hand. Jiro explained to us how in ceremonies such as weddings, each diner should have their own lobster, and, although servings may not have always been so diminutive, the lobsters most sought these days tended to be small simply because of their high cost. The red rock lobster *Jasus edwardsii* from New Zealand and Australia is similar to the Japanese spiny lobster—indeed, among the imported species, *Jasus* lobsters are preferred because their colour and taste are so comparable. But Japanese consumers certainly know their stuff and the price for the imports is significantly lower. Packhorse realise even less on this market, for they are green or brown before cooking, rather than red. And generally they're just too large.

Japanese spiny lobsters are highly revered for ceremonial occasions. Small ones on offer at fish markets help to make affordable the custom of each guest being served their own lobster.

Until the early 1990s, most packhorse in New Zealand were either cooked whole in large boilers or tailed before being exported. In tailing, the lobster tail is held in one hand and bent up, the top of the carapace in the other hand. The lobster is then slid onto a horizontally held, 25-centimetre-long stainless steel blade that is slightly curved in cross section and so resembles a long, straight, shiny tongue. The blade rudely enters the animal at the back and base of its body. Once fully engaged, the body is partially rotated back and forth and at the same time pulled away from the tail. This leaves the complete tail, with its large bulge of muscle that is normally safely hidden within the body cavity, in the butcher's hand. A blob of digestive gland (also known as 'mustard' because of its colour) drops from the body. Disdainfully left on the factory floor to be later hosed into the drains, this material is highly esteemed elsewhere. In Japan it is known as 'miso' (not to be confused with soy or grain-based miso) and commands a value-per-kilogram many times that of the lobster itself. Mature females may also have a pair of long, dark-brick-red sacs of developing eggs that emerge adhering to the tails. Just before egg extrusion these sacs become greatly engorged, extending

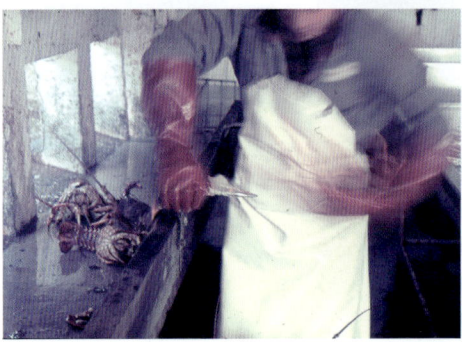

From earlier times: tailing live rock lobsters (left). Those not tailed were boiled (above).

through almost all of the spare body-cavity space and even forward to the base of the antennae, as we saw in Chapter 5. Again, in certain places this is a delicacy, which cannot be said of the eggs once they are outside, attached to the tail.

The rear end of the tail is then slid onto a much smaller, circular blade that cuts the tissue securing the anus to the exoskeleton. This frees the last part of the digestive tract, which is drawn or air-blown out of the tail as a long white tube. The tail is then held in chilled seawater before being graded, packed and frozen. Meanwhile, the body, with all its five pairs of legs, its feelers and its other appendages now engaged in uncontrolled stammers of movement, is consigned to some temporary repository. Later it will be delivered to a destination which varies according to demand: cooked and marketed, or just trucked to the local dump.

All the while the stainless-steel tailing table and surrounds become covered in a slightly gooey fluid—blood. It is clear to start with but becomes a light metallic blue as it congeals because copper, rather than iron, binds to the oxygen.

Today most spiny lobsters are marketed live and intact, which is all round more desirable. There is far too much of good flavour and texture within the body to waste it by throwing it away. When lobsters are tailed, it is done much more humanely, as described later.

Spiny lobsters are an international commodity—a pricey one at that—and naturally, fishers seek what the global market will pay. In addition to the United States, the major spiny lobster importing nations are in Asia and Europe. New Zealand packhorse are exported mainly to China and Hong Kong, the rest being delivered live to a fashionable domestic restaurant market.

And it's fortunate that the spiny lobster is a marine creature that, if looked after, survives quite happily out of water for many hours. Fishers may receive twice as much for a lobster that can be marketed live compared with one the same size which has had to be tailed because it is missing limbs or is weak. Accordingly, in the interests of bringing top-quality fare to market and simply being humane, codes of practice guide the way that lobsters in New Zealand are handled 'from water to waiter'. Key words include 'cover, chill, aerate, keep damp'. On the packhorse boats I went on the lobsters were immediately dropped into a large on-deck tank that was being constantly supplied with fresh seawater. At night they were transferred to holding pots—large pots without funnels marked by midget buoys that their owners hoped no one else would find.

At dockside, the lobsters are typically transferred to plastic fish bins that can be stacked without crushing their incumbents. But once at the packing factory, individual lobsters—rather than entire boxes of them—become the focus. Air exposure leads to carbon dioxide accumulating in the blood and tissues. It brings on acidosis, the lobster becoming lethargic, the tail limp, and limb movements weak.[1] Any acidosis which has already taken place at sea and on the road must be reversed before the lobsters are packed. Those that will not survive live storage and transport to market are separated out for tailing. The packers don't want the bad look of a dying or dead lobster among their vital ones. The lively ones are now in for a few days of special care.

Spiny lobsters are held live before marketing for all manner of reasons: amassing sufficient quantity, marketing strategy, allowing the lobsters to grow more, and so on. And the trick is to ensure that the quality and characteristics of the water they are held in are up to standard. High water temperatures are normally more injurious than low ones, probably because they lead to malfunction of the immune system. On the other hand, low temperatures slow enzyme reactions and the lobsters become comatose.[1] Spiny lobsters can adjust to the ambient salinity over only a limited range. Nevertheless, most species will put up with gradual reductions in salinity to surprisingly low values—in the order of 20 parts per thousand below

It's hard to see how this heap of large red rock lobsters can be kept in good condition to reach any dinner table, particularly as it is doubtful there is refrigeration handy.

Commercial fishers like Nat Davey (left) keep their packhorse live in fine condition in holding pots until they are ready to land them. These lobsters' ultimate destination is the lucrative live market, where very little is wasted. At $90 per kilogram (right), not everyone will be able to afford these live packhorse at this Auckland fish market—no matter how much their favourite seafood they might be.

that of the water they normally occupy. This means that live-holding plants can work perfectly well with waters that are somewhat lower in salt content than the sea itself.

Shortage of oxygen in the water is the main killer in live-holding plants. Spiny lobsters are referred to as oxygen regulators, being able to maintain a relatively normal uptake of oxygen under conditions of high or low oxygen availability, but this is true only to around 30 per cent saturation. Indeed, the recommended minimum dissolved oxygen level in which to hold spiny lobsters is at least 70 per cent saturation.[2] And there is less oxygen present in seawater the higher the temperature and the higher the salinity. An unexpected problem is too much air in the water—supersaturation—which can lead to potentially fatal gas bubble disease. It can be seen through the transparent underside of the spiny lobster's tail as bubbles in the tissue. The condition most often results from air leakage on the intake side of the water pump and is remedied by mechanically mixing the water with air through such means as vigorously spraying the water before it enters the tanks.

The next most threatening condition to spiny lobsters is too much ammonia (or its break-down products) in the water. High levels of ammonia—the result of the breakdown of proteins—can accumulate in the holding water and then in the lobster's blood, damage cells and eventually kill. Meanwhile, bacteria in the water convert ammonia to nitrite, then to nitrate. Both are lethal at high concentrations. (Being a bit technical for a moment, ammonia should not exceed 2 milligrams per litre; nitrite 5 milligrams per litre; and nitrate 100 milligrams per litre.[2] Off-the-shelf kits are readily available to make the measurements.)

The methods used to pack and transport live spiny lobsters are essentially the same the world over. As described by Richard Stevens of the Western Australian Fishing Industry Council and Daryl Sykes of the New Zealand Rock Lobster Industry Council,[3] live spiny lobsters are typically held at ambient temperature for 3 days in flow-through or recirculating seawater holding tanks. This allows them to purge their gut contents, and for the weak ones to be identified and culled. Also removed are soft- and new-shell lobsters, neither of which survive well because moulting has depleted their energy reserves.

The lobsters are immobilised by lowering the water temperature, often down to 3–4°C over the final 12 hours. They are then firmly packed in 10-kilogram lots in light and insulating polystyrene boxes, with wood shavings or expanded polystyrene separating individuals to keep them from damaging each other. (Sawdust is not appreciated for it takes too much water and time to clean off at the other end.) The antennae may be deliberately broken off near their base and placed in the carton alongside the lobsters, to be refitted later. (A stockpile of antennae from tailed individuals provides cover for the occasional live lobster missing its own.) A plastic bottle or two of frozen seawater, or a special cooling

Packing live red rock lobsters for export. They will readily survive out of the water for 30 hours.

bag, is added to keep temperatures low. Ventilation is by means of small holes in the box cover. In this way, spiny lobsters easily survive for at least 30 hours. Transferred to holding tanks at their destination, they are then kept alive until selected by their ultimate predator.

Because many of the markets are distant from clean seawater, the holding tanks at the lobsters' destination may be composed of quite-expensive concocted seawater (salt added to freshwater). Often, then, the frozen seawater that had been packed with the lobsters to keep them cool during their journey is appropriated to top up these tanks. A worrisome implication of the live transport of spiny lobsters is the massive quantity of expanded polystyrene that accumulates at the markets. For example, New Zealand exports about 2500 tonnes of red rock lobsters *Jasus edwardsii* each year, which can involve 250 000 polystyrene boxes. Expanded polystyrene—which is 90 per cent air—is not a high priority target for recycling, and most of it is dumped.

Most—but not all—spiny lobsters are exported in this manner, alive. For whole raw spiny lobster, the lobsters are anaesthetised and suffocated in iced freshwater, graded, wrapped, packed, blast frozen to –40°C, and then stored below –25°C.[3] For a long commercial life, freezing must be rapid because enzymes in the internal organs begin to break them down the moment the lobster dies. For whole cooked spiny lobsters, the animals are boiled in salted water for between 8 and 20 minutes depending on size (temperature at the centre of the thickest part of the tail needs to reach 70°C), then rapidly chilled.

For spiny lobster tails, the lobsters are anaesthetised and suffocated in iced freshwater, the tail removed, and the hindgut vacuum-extracted. This leaves the bodies, which are also graded, packed and frozen. Another byproduct of tailing is paste, where the bodies are finely macerated and frozen into blocks.[3]

Spiny lobsters are widely revered for their taste, the tail muscle being rich in free amino acids (particularly glycine) which give it a characteristic sweetness.[4] Seawater concentration influences the levels of free amino acids and therefore flavour, which can, it is said, be enhanced by immersing the lobsters in concentrated seawater for short periods. This may also explain why decapods from deep waters, which are cool and salty, are often particularly sweet and flavoursome. But the taste-active components are also influenced by diet, growth, moult stage, maturity, season and freshness—only a couple of which can be controlled.

The Asian spiny lobster market is quite a finicky one. For example, the lobster's colour before cooking greatly affects its value—usually the darker the red the higher the price. Markdown because of poor colour equates to losses of AUD$7 million for Tasmanian *Jasus edwardsii*, the differential being greatest in summer, near the new year.[5] Japanese people are particularly fond of spiny lobsters, along with shrimps and prawns, not just because of their flavour but also the striking red they become *after* cooking. Their own spiny lobster, in particular, has become a symbolic decoration at New Year's Day feasts, and the main and indispensable dish at wedding ceremonies—where everyone must be served their own lobster. The pair of long antennae symbolises happiness and longevity; to sever them is to invite a break up. Recent great wealth in parts of China has led to rapid expansion of the market there for spiny lobsters. There are several fisheries along the coast of China but nowhere near enough spiny lobsters are caught to satisfy the country's demand; most are imported from Australia and New Zealand. And it is in this market that packhorse find a special place. They are excellent centrepieces for banquets, their large size and good flavour bringing admiration and esteem to the host. And importantly they lack pincers—once again to do with the severing of relationships.

HOW TO PREPARE AND POLISH OFF YOUR LOBSTER

It's widely held that, whatever method you use to despatch your lobster, you should chill it thoroughly first to induce anaesthesia and render slaughter more humane.[7] In cold-blooded creatures such as spiny lobsters, chilling helps reduce nerve function and metabolic activity. The lobster can be placed in a slurry of ice in salt water, or into a refrigerator set between 2 and 4°C. When fully chilled it stops moving, becomes limp, and does not respond to being handled. Chilling to temperatures above 4°C does not guarantee complete insensibility.[8]

Splitting or spiking are considered to be the most humane methods of killing lobsters,[8] but be aware that each leads to the severing of the stomach and release of stomach acid—a sour point if you intend later to suck the goodness from inside the body. Splitting the lobster along its length severs the nerves linking the ganglionic chains that run the length of its body, and is thought to prevent it feeling anything. Head spiking (inserting a blade between the eyes) injures the main ganglia at the front of the animal. In contrast, tailing and chest spiking (pushing a knife into the chest wall from the underside) do not cause immediate unconsciousness in unchilled lobsters. For your sake and the lobster's, forget about putting it directly into boiling water. Inhumane too is placing spiny lobsters in unsalted tap water. This causes struggling, severe osmotic stress and slow drowning—and can also make the meat soft because it absorbs water.[8]

For the innermost tissues to reach 70°C, a smallish lobster will take 10 minutes in already-boiling water. Then cool it in running cold water. Lobsters split and barbecued are not for me: I find the smell and flavour of blackened shell permeating and offensive. On the other hand, serving the lobster chilled, sashimi-style, dipped in soya sauce with wasabi, is certainly worth trying. The texture and flavour are very different to those of cooked meat.

Over to you.

However you chose to enjoy your packhorse tail, cooked or raw, don't forget about all the goodies still within the body.

EPILOGUE

Packhorse in New Zealand. They have more than tolerably well represented the doings of the intriguing family of marine creatures which has been the focus of this book. Spiny lobsters have been my subjects, packhorse the linchpin.

I end by reflecting on the future of packhorse in this country. A distinctive spiny lobster in many ways—the most migratory, the most fecund, found only in this part of the world, with a tiny fishery focused on the main breeding grounds—its total allowable commercial catch (TACC) has never been reached. According to the *Best Fish Guide 2009–10*,[1] packhorse is ranked as 'Amber—Concerns', one up from 'Red—Species to avoid'. Their reasons for this ranking? There is little basic biological information on the species, and no formal stock assessment or directed research; the cause of declines in their reported catch is unknown; and potting them can harm sensitive seabed habitats.

None of these concerns is unreasonable. There is no doubt that over the years very large catches of packhorse have been taken from quite small areas of New Zealand. Most people in the industry would agree that the stock has been hit for six. Recent annual landings average well below the TACC of 40.3 tonnes, nothing like the 100–200 tonnes per year that were caught in the 1960s and 1970s.

Perhaps not surprisingly, then, I started off expecting that by the end of these writings I would recommend closing down the fishery. Someone ought to buy out and lock up the quota.

But there are other ways of looking at this fishery. Tiny though it may be, it is nevertheless important locally. It is essentially protected from recruitment overfishing by the fact that the adults are so migratory and therefore difficult to catch, with harvests restricted to a very isolated and exposed place where fishing is difficult and can be dangerous. And there are several possible reasons for the

TACC remaining uncaught, and for the fluctuations in annual catch. Firstly, the TACC may be just too high given the stock size. No serious research has been put into determining what the TACC should be. Secondly, the large number of small quota holders, many of them retaining quota just to cover any packhorse bycatch, may not be catching their full entitlement each year. Thirdly, some of the handful of major players may in some years be targeting their fishing towards species other than packhorse. Fourthly, there are sure to be natural variations in packhorse abundance, which, together with the weather in some years being just too wild for fishing those exposed and remote northern grounds, contrive to make annual landings jump all over the place.

What does seem clear is that the packhorse fishery in New Zealand is one likely to benefit from shrewder management. This *might* include a smaller minimum legal size, possibly combined with a maximum legal size to protect the larger breeders—as is the case in New South Wales. It is a fishery which has every chance of being able to produce significantly more revenue than it currently does, because the smaller packhorse are so valuable.

But perhaps the foremost issue is not so much management of packhorse in the main fishing grounds of the Far North, but the continued fish-down of the last small bastions of large packhorse along the east coast of the North Island. Practically, the east coast is the only place you and I could ever experience a large packhorse. The Far North grounds are just too inaccessible. Could packhorse—except in the Far North—be made illegal to land, with the commercial quota being bought up to compensate? Migrating packhorse would then be able to make it back to their breeding grounds, unmolested; those few that decide to stop off will again gradually accumulate along the east coast. Few commercial fishers would be affected, and then only a little because so few legal packhorse are taken south of North Cape.

The upshot would be that divers would be more likely to encounter, with wonder, a really large spiny lobster. The largest on Earth.

Probably the greatest threat to packhorse today is that they don't really appear on the radar screen, there being little dialogue among the movers and shakers as to how they should be managed. They are considered too insignificant, too enigmatic—simply not worth the effort. Maybe now is the time to discuss this fishery more seriously. Like when a plug is accidentally knocked out of a sink of very hot water, we somehow need to get our hand in there to stem the flow before all the water has drained. And, to my mind, it is those last little pockets of giant packhorse still to be found south of North Cape that will be the first treasures to slip down the plughole.

I remain as inquisitive about packhorse as the demoiselle at the bottom of this photo appears to be. Primitive yet majestic. A drifter of the high seas, yet able to navigate with surety hundreds of kilometres over unknown terrain. Relatively rarely encountered in the wild, yet coping well enough after half a century of fishing. A hardy creature at the forefront of aquaculture. A survivor, we hope . . .

REFERENCES

1. WIDE WANDERINGS

1 Nichol, S.L.; Regnauld, H.; Goff, J.R. 2004: Sedimentary evidence for tsunami on the northeast coast of New Zealand. *Géomorphologie: relief, processus, environnement 1*: 35–44.

2 Hayward, B.W. 2009: Personal communication. (Geomarine Research, 49 Swainston Rd, St Johns, Auckland, New Zealand.

3 Ingram, C.W.N.; Wheatley, P.O.; Diggle, L.; Diggle, E.; Gordon, K.R. 2007: *New Zealand shipwrecks: over 200 years of disasters at sea*. Hodder Moa, Auckland, New Zealand.

4 Booth, J.D. 1979: North Cape—a 'nursery area' for the packhorse rock lobster, *Jasus verreauxi* (Decapoda: Palinuridae). *New Zealand Journal of Marine and Freshwater Research 13*: 521–528.

5 Booth, J.D. 1986: Recruitment of packhorse rock lobster *Jasus verreauxi* in New Zealand. *Canadian Journal of Fisheries and Aquatic Sciences 43*: 2212–2220.

6 Meek, A. 1915: Migrations in the sea. *Nature 95*: 231.

7 Phillips, B.F.; Chubb, C.F.; Melville-Smith, R. 2000: The status of Australia's rock lobster fisheries. *In*: *Spiny lobsters: fisheries and culture*. Phillips, B.F; Kittaka, J. (eds) pp 45–77. Fishing News Books, Oxford.

8 Street, R.J. 1995: Rock lobster migrations in southern New Zealand. *Seafood New Zealand August*: 19–21.

9 Kendrick, T.H.; Bentley, N. 2003: Movements of rock lobsters (*Jasus edwardsii*) tagged by commercial fishers around the coast of New Zealand from 1993. New Zealand Fisheries Assessment Report 2003/55.

10 Boles, L.C.; Lohmann, K.J. 2003: True navigation and magnetic maps in spiny lobsters. *Nature 421*: 60–63.

11 Lohmann, K.J.; Pentcheff, N.D.; Nevitt, G.A.; Stetten, G.D.; Zimmer-Faust, R.K.; Jarrard, H.E.; Boles, L.C. 1995: Magnetic orientation of spiny lobsters in the ocean:

Experiments with undersea coil systems. *Journal of Experimental Biology 198*: 2041–2048.

12 Jeffs, A.G., Tolimieri, N., Haine, O.; Montgomery, J.C. 2003: Crabs on cue for the coast: the use of underwater sound for orientation by pelagic crab stages. *Marine and Freshwater Research 54*: 841–845.

13 MacDiarmid, A.B. 1991: Seasonal changes in depth distribution, sex ratio and size frequency of spiny lobster *Jasus edwardsii* on a coastal reef in northern New Zealand. *Marine Ecology Progress Series 70*: 129–141.

14 Cockcroft, A.C. 2001: *Jasus lalandii* 'walkouts' or mass strandings in South Africa during the 1990s: an overview. *Marine and Freshwater Research 52*: 1085–1094.

15 Anon 2006: *African Journal of Marine Science 28(2)*: cover.

2. SO, JUST WHAT ARE PACKHORSE?

1 Pike, R.B. 1969: A case study in research: crayfish. *In*: *Fisheries and New Zealand.* Slack, E.B. (ed) pp 95–108. Proceedings of a seminar on fisheries development in New Zealand. Department of University Extension, Victoria University of Wellington.

2 Holthuis, L.B. 1991: *Marine lobsters of the world. An annotated and illustrated catalogue of species of interest to fisheries known to date.* FAO Fisheries Synopsis 125.

3 Anon 2008: Henri Milne-Edwards in Wikipedia. http://en.wikipedia.org/wiki/Henri_Milne-Edwards.

4 Anon 2009: Jules Verreaux in Wikipedia. http://en.wikipedia.org/wiki/Jules_Verreaux.

5 Milne Edwards, H. 1851: Observations sur le squelette tégumentaire des Crustacés décapodes, et sur la morphologie de ces animaux. *Annales des Sciences Naturelles. La Zoologie 16*: 221–291.

6 Heller, C. 1862: Beiträge zur näheren Kenntnis der Macrouren. *Sitzungsberichte der Akademie der Wissenschaften, mathematisch-naturwissenschaftliche Classe, Wien 45(1)*: 389–426.

7 Kirk, T.W. 1880: Description of a new species of *Palinurus*. *Transactions and Proceedings of the New Zealand Institute 12*: 313–314.

8 Archey, G. 1916: Notes on the marine crayfish of New Zealand. *Transactions and Proceedings of the New Zealand Institute 48*: 396–406.

9 Buchanan, John. *In*: *An encyclopaedia of New Zealand.* Originally published in 1966, McLintock, A.H. (ed). Te Ara — The encyclopedia of New Zealand www.teara.govt.nz.

10 Haswell, W.A. 1882: *Catalogue of the Australian stalk- and sessile-eyed Crustacea.* The Australian Museum, Sydney.

11 Parker, T.J. 1883: On the structure of the head in '*Palinurus*,' with special reference to the classification of the genus. *Nature* 29: 189–190.

12 Booth, J.; Webber, R. 2001: All the pretty lobsters – Part I. *Seafood New Zealand December*: 20-23.

13 Booth, J.D. 2006: *Jasus* species. *In: Lobsters: biology, management, aquaculture and fisheries.* Phillips, B.F. (ed) pp 340–358. Blackwell Publishing Ltd, Oxford.

14 Brasher, D.J.; Ovenden, J.R.; Booth, J.D.; White, R.W.G. 1992: Genetic subdivision of Australian and New Zealand populations of *Jasus verreauxi* (Decapoda: Palinuridae)—preliminary evidence from the mitochondrial genome. *New Zealand Journal of Marine and Freshwater Research 26*: 53–58.

15 Palero, F.; Crandall, K.A.; Abelló, P.; Macpherson, E.; Pascual, M. 2009: Phylogenetic relationships between spiny, slipper and coral lobsters (Crustacea, Decapoda, Achelata). *Molecular Phylogenetics and Evolution 50*: 152–162.

16 Chan, T.-Y. 2010: Annotated checklist of the world's marine lobsters (Crustacea: Decapoda: Astacidea, Glypheidea, Achelata, Polychelida). *The Raffles Bulletin of Zoology Supplement 23*: 153–181.

17 Glaessner, M.F. 1960: The fossil decapod Crustacea of New Zealand and the evolution of the Order Decapoda. New Zealand Department of Scientific and Industrial Research Paleontological Bulletin 31.

18 Dixon, C.J.; Ahyong, S.T.; Schram, F.R. 2003: A new hypothesis of decapod phylogeny. *Crustaceana 76*: 935–975.

19 Patek, S.N.; Feldmann, R.M.; Porter, M.; Tshudy, D. 2006: Phylogeny and evolution. *In: Lobsters: biology, management, aquaculture and fisheries.* Phillips, B.F. (ed) pp 113–145. Blackwell Publishing Ltd, Oxford.

20 Campbell, H.; Hutching, G. 2007: *In search of ancient New Zealand*. Penguin Books, North Shore.

21 George, R.W. 2006: Tethys origin and subsequent radiation of the spiny lobsters (Palinuridae). *Crustaceana 79*: 397–422.

22 George, R.W.; Main, A.R. 1967: The evolution of spiny lobsters (Palinuridae): a study of evolution in the marine environment. *Evolution 21*: 803–820.

23 George, R.W. 2005: Evolution of life cycles, including migration, in spiny lobsters (Palinuridae). *New Zealand Journal of Marine and Freshwater Research 39*: 503–514.

24 George, R.W. 1997: Tectonic plate movements and the evolution of *Jasus* and *Panulirus* spiny lobsters (Palinuridae). *Marine and Freshwater Research 48*: 1121–1130.

25 Pollock, D.E. 1990: Palaeoceanography and speciation in the spiny lobster genus *Jasus. Bulletin of Marine Science 46*: 387–405.

26 Palero, F.; Crandall, K.A.; Abelló, P.; Macpherson, E.; Pascual, M. 2009: Phylogenetic relationships between spiny, slipper and coral lobsters (Crustacea, Decapoda, Achelata). *Molecular Phylogenetics and Evolution 50*: 152–162.

3. PRESENCE AND ABSENCE

1 Booth, J.D. 1986: Recruitment of packhorse rock lobster *Jasus verreauxi* in New Zealand. *Canadian Journal of Fisheries and Aquatic Sciences 43*: 2212–2220.
2 Hicks, D.M.; Gomez, B.; Trustrum, N.A. 2004: Event suspended sediment characteristics and the generation of hyperpycnal plumes at river mouths: East coast continental margin, North Island, New Zealand. *Journal of Geology 112*: 471–485.
3 Holthuis, L.B. 1991: *Marine lobsters of the world. An annotated and illustrated catalogue of species of interest to fisheries known to date.* FAO Fisheries Synopsis 125.
4 Montgomery, S.S.; Craig, J.R. 2005: Distribution and abundance of recruits of the eastern rock lobster (*Jasus verreauxi*) along the coast of New South Wales, Australia. *New Zealand Journal of Marine and Freshwater Research 39*: 619–628.
5 Anon 2011: Ocean in Wikipedia. http://en.wikipedia.org/wiki/Ocean
6 Mironov, A.N.; Molodtsova, T.N.; Parin, N.V. 2008: Soviet and Russian studies on seamount biology. www.isa.org.jm/files/documents/EN
7 Heydorn, A.E.F. 1969: The South Atlantic rock lobster *Jasus tristani* at Vema Seamount, Gough Island and Tristan da Cunha. Republic of South Africa Department of Industries Division of Sea Fisheries Investigational Report 73.
8 Eade, J.V. 1976: Geological notes on the Southwest Pacific Basin in the area of Wachusett Reef and Maria Theresa Reef. NZOI Oceanographic Summary 9.
9 Chadderton, J. 2008: Personal communication.
10 Booth, J.D.; Street, R.J.; Smith, P.J. 1990: Systematic status of the rock lobsters *Jasus edwardsii* from New Zealand and *J. novaehollandiae* from Australia. *New Zealand Journal of Marine and Freshwater Research 24*: 239–249.
11 Parin, N.V.; Mironov, A.N.; Nesis, K.N. 1997: Biology of the Nazca and Sala y Gómez submarine ridges, an outpost of the Indo-west Pacific fauna in the eastern Pacific Ocean: composition and distribution of the fauna, its communities and history. *In*: *The biogeography of the oceans.* Blaxter, J.H.S.; Southward, A.J.; Gebruk, A.V.; Southward, E.C.; Tyler, P.A. (eds) pp 145–242. Advances in Marine biology. Academic Press, London.
12 Webber, W.R.; Booth, J.D. 1988: *Projasus parkeri* (Stebbing, 1902) (Crustacea, Decapoda, Palinuridae) in New Zealand and description of a *Projasus* puerulus from Australia. *National Museum of New Zealand Records 3*: 81-92.

13 Groeneveld, J.C.; Goni, R.; Latrouite, D. 2006: *Palinurus* species. *In*: *Lobsters: biology, management, aquaculture and fisheries.* Phillips, B.F. (ed) pp 385–411. Blackwell Publishing Ltd, Oxford.

14 Groeneveld, J.; Griffiths, C.; Matthee, C. 2008: Meet *Palinurus barbarae* – a new spiny lobster species from a submerged seamount! *The Lobster Newsletter 21(1)*: 2–3.

15 Booth, J.D. 2006: *Jasus* species. *In*: *Lobsters: biology, management, aquaculture and fisheries.* Phillips, B.F. (ed) pp 340–358. Blackwell Publishing Ltd, Oxford.

16 *The Evening Post* 1986: Kiwis stick with shelled Raider. 16 October 1986.

17 *The Evening Post* 1986: NZ relatives await news of missing crew. 14 October 1986.

18 *The West Australian* 1987: France: Trawler was not fishing. 19 February 1987.

19 Webber, W.R.; Booth, J.D. 1995: A new species of *Jasus* (Crustacea: Decapoda: Palinuridae) from the eastern South Pacific Ocean. *New Zealand Journal of Marine and Freshwater Research 29*: 613–622.

20 Phillips, B.F.; Melville-Smith, R. 2006: *Panulirus* species. *In*: *Lobsters: biology, management, aquaculture and fisheries.* Phillips, B.F. (ed) pp 359–384. Blackwell Publishing Ltd, Oxford.

21 Sekiguchi, H.; George, R.W. 2005: Description of *Panulirus brunneiflagellum* new species with notes on its biology, evolution, and fisheries. *New Zealand Journal of Marine and Freshwater Research 39*: 563–570.

22 Chan, T-Y.; Ng, P.K.L. 2001: On the nomenclature of the commercially important spiny lobsters *Panulirus longipes femoristriga* (Von Martens, 1872), *P. bispinosus* Borradaile, 1899, and *P. albiflagellum* Chan & Chu, 1996 (Decapoda, Palinuridae). *Crustaceana 74*: 123-127.

23 George, R.W.; Main, A.R. 1967: The evolution of spiny lobsters (Palinuridae): a study of evolution in the marine environment. *Evolution 21*: 803–820.

24 Newman, W.A. 1991: Origins of Southern Hemisphere endemism, especially among marine Crustacea. *Memoirs of the Queensland Museum 31*: 51–76.

25 Butler, M.J.; Steneck, R.S.; Herrnkind, W.F. 2006: Juvenile and adult ecology. *In*: *Lobsters: biology, management, aquaculture and fisheries.* Phillips, B.F. (ed) pp 263–309. Blackwell Publishing Ltd, Oxford.

26 Radhakrishnan, E.V. 1995: Lobster fisheries of India. *The Lobster Newsletter 8(1)*: 1, 12–13.

27 Talley, L.D. 2007: *Hydrographic atlas of the World Ocean Circulation Experiment (WOCE). Volume 2: Pacific Ocean.* National Oceanography Centre, Southampton.

28 Chan, T.-Y. 2010: Annotated checklist of the world's marine lobsters (Crustacea: Decapoda: Astacidea, Glypheidea, Achelata, Polychelida). *The Raffles Bulletin of Zoology Supplement 23*: 153–181.

29 Poupin, J. 2010: Biodiversité de l'Indo-Pacifique tropical français: 2514 espèces de crustacés décapodes et stomatopodes. *Rapport Scientifique Institut de Recherche de l'Ecole Navale*, Octobre 2010.

30 Felder, D.L.; Álvarez, F.; Goy, J.W.; Lemaitre, R. 2009: Decapoda (Crustacea) of the Gulf of Mexico, with comments on the Amphionidacea. *In: Gulf of Mexico origin, waters, and biota. Volume 1, Biodiversity.* Felder, D.L.; Camp, D.K. (eds) pp 1019–1104. Texas A&M University Press.

31 Chan, T.-Y.; Yu, H.-P. 1995: The rare lobster genus *Palinustus* A. Milne Edwards, 1880 (Decapoda: Palinuridae), with description of a new species. *Journal of Crustacean Biology 15*: 376–394.

32 Poupin, J. 1994: The genus *Justitia* Holthuis, 1946, with the description of *J. chani* and *J. vericeli* spp. nov. (Crustacea: Decapoda: Palinuridea). *Journal of Taiwan Museum 47*: 37–56.

33 Wowor, D. 1999: The spear lobster, *Linuparus somniosus* Berry & George, 1972 (Decapoda, Palinuridae) in Indonesia. *Crustaceana 72*: 673–684.

34 Tomoyuki, K.; Atsushi, K.; Keiichi, S. 2005: New record of the spear lobster *Linuparus sordidus* Bruce (Crustacea: Decapoda: Palinuridae) from Japanese waters. *Biological Magazine Okinawa 43*: 15–20.

35 Chan, T.-Y. 1997: Crustacea Decapoda: Palinuridae, Scyllaridae and Nephropidae collected in Indonesia by the KARUBAR Cruise, with an identification key for the species of *Metanephrops*. *In: Résultats des Campagnes MUSORSTOM, Volume 16.* Crosnier, A.; Bouchet, P. (eds). *Mémoires Muséum National d'Histoire Naturelle* 172: 409–431.

36 Haddy, J.A.; Roy, D.P.; Courtney, A.J. 2003: The fishery and reproductive biology of barking crayfish, *Linuparus trigonus* (von Siebold, 1824) along Queensland's east coast. *Crustaceana 76*: 1189–1200.

37 Davie, P.J.F. 2002: Crustacea: Malacostraca: Phyllocarida, Hoplocarida, Eucarida (Part 1). *In: Zoological catalogue of Australia 19.3A.* Wells, A.; Houston, W.W.K. (eds). CSIRO Publishing.

38 O'Malley, J.M. 2009: Spatial and temporal variability in growth of Hawaiian spiny lobsters in the Northwestern Hawaiian Islands. *Marine and Coastal Fisheries: Dynamics, Management, and Ecosystem Science 1*: 325–342.

39 Domingue, G. 1996: Spiny lobster fishery around the Mahe Plateau: results and analysis presentation, November 1995–January 1996. Seychelles Fishing Authority Technical Report 40.

40 Booth, J.; Webber, R. 2001: All the pretty lobsters – Part I. *Seafood New Zealand December*: 20-23.

41 Bliss, D. 1982: *Shrimps, lobsters and crabs.* New Century Publishers, Piscataway, New Jersey.

42 Vianna, M.L. 1986: On the ecology and intraspecific variation in the spiny lobster *Panulirus echinatus* Smith. 1869. (Decapoda, Palinuridae) from Brazil. *Crustaceana 51*: 25–37.

43 Pollock, D.E. 1992: Palaeoceanography and speciation in the spiny lobster genus *Panulirus* in the Indo-Pacific. *Bulletin of Marine Science 51*: 135–146.

44 Branch, G.M.; Griffiths, C.L.; Branch, M.L.; Beckley, L.E. 1994: *Two oceans. A guide to the marine life of Southern Africa.* David Philip, Cape Town.

45 Phillips, B.F.; Melville-Smith, R. 2006: *Panulirus* species. *In*: Lobsters: biology, management, aquaculture and fisheries. Phillips, B.F. (ed) pp 359–384. Blackwell Publishing Ltd, Oxford.

46 Database of Central Pacific Crustacea (Decapoda and Stomatopoda). http://decapoda.free.fr/fiche.php?sp=558. 17 March 2011.

4. BIZARRE BEGINNINGS

1 McWilliam, P.S.; Phillips, B.F. 1987: Distinguishing the phyllosoma larvae of rock lobster species of the genus *Jasus* (Decapoda, Palinuridae) in the waters of Australia and New Zealand. *Crustaceana 52*: 1–24.

2 Kensler, C.B. 1967: Fecundity in the marine spiny lobster *Jasus verreauxi* (H. Milne Edwards) (Crustacea: Decapoda: Palinuridae). *New Zealand Journal of Marine and Freshwater Research 1*: 143–155.

3 Kittaka, J.; Ono, K.; Booth, J.D. 1997: Complete development of the green rock lobster, *Jasus verreauxi* from egg to juvenile. *Bulletin of Marine Science 61*: 57–71.

4 Montgomery, S.S.; Kittaka, J. 1994: Occurrence of pueruli of *Jasus verreauxi* (H. Milne Edwards, 1851) (Decapoda, Palinuridae) in waters off Cronulla, New South Wales, Australia. *Crustaceana 67*: 65–70.

5 Cunningham, J.T. 1891: On the development of *Palinurus vulgaris*, the rock lobster or sea crayfish. *Journal of the Marine Biological Association of the United Kingdom 2*: 141–150.

6 Lesser, J.H.R. 1978: Phyllosoma larvae of *Jasus edwardsii* (Hutton) (Crustacea: Decapoda: Palinuridae) and their distribution off the east coast of the North Island, New Zealand. *New Zealand Journal of Marine and Freshwater Research 12*: 357–370.

7 McWilliam, P.S.; Phillips, B.F. 1992: The final and subfinal larval stages of *Panulirus polyphagus* (Herbst) and the final stage of *Panulirus ornatus* (Fabricius), with a review

of late-stage larvae of the *Panulirus homarus* larval complex (Decapoda, Palinuridae). *Crustaceana 62*: 249–272.

8   Johnson, M.W.; Robertson, P.B. 1970: On the phyllosoma larvae of the genus *Justitia* (Decapoda, Palinuridae). *Crustaceana 18*: 283–292.

9   Bouvier, M.E.-L. 1914: Recherches sur le développement post-embryonnaire de la langouste commune (*Palinurus vulgaris*). *Journal of the Marine Biological Association of the United Kingdom 10*: 179–193.

10 Palero, F.; Guerao, G.; Clark, P.F.; Abelló, P. 2010: Final-stage phyllosoma of *Palinustus* A. Milne-Edwards, 1880 (Crustacea: Decapoda: Achelata: Palinuridae)—The first complete description. *Zootaxa 2403*: 42–58.

11 Johnson, M.W. 1968: On phyllamphion larvae from the Hawaiian Islands and the South China Sea. *Crustaceana* Supplement 2: 38–46.

12 Sekiguchi, H.; Booth, J.D.; Kittaka, J. 1996: Phyllosoma larva of *Puerulus angulatus* (Bate, 1888) (Decapoda: Palinuridae) from Tongan waters. *New Zealand Journal of Marine and Freshwater Research 30*: 407-411.

13 Sims, H.W. 1966: The phyllosoma larvae of the spiny lobster *Palinurellus gundlachi* von Martens (Decapoda, Palinuridae). *Crustaceana 11*: 201–215.

14 Sekiguchi, H.; Booth, J.D.; Webber, W.R. 2007: Early life histories of slipper lobsters. In: *The biology and fisheries of the slipper lobster*. Lavalli, K.L.; Spanier, E. (eds) pp 69–90. CRC Press, London.

15 Kittaka, J. 1994: Culture of phyllosomas of spiny lobster and its application to studies of larval recruitment and aquaculture. *Crustaceana 66*: 258–70.

16 Cox, S.L.; Johnston, D.J. 2003: Feeding biology of spiny lobster larvae and implications for culture. *Reviews in Fisheries Science 11*: 89–106.

17 Jeffs, A.G.; Phleger, C.F.; Nelson, M.M.; Mooney, B.D.; Nichols, P.D. 2002: Marked depletion of polar lipid and non-essential fatty acids following settlement by post-larvae of the spiny lobster *Jasus verreauxi*. *Comparative Biochemistry and Physiology 131A*: 305–311.

18 Hayward, B.W. 2009: Personal communication. (Geomarine Research, 49 Swainston Rd, St Johns, Auckland, New Zealand.)

19 Cryer, M.; O'Shea, S.; Gordon, D.; Kelly, M.; Drury, J.; Morrison, M.; Hill, A.; Saunders, H.; Shankar, U.; Wilkinson, M.; Foster, G. 2000: Distribution and structure of benthic invertebrate communities between North Cape and Cape Reinga. Final Research Report for Ministry of Fisheries Research Project ENV9805.

20 Bradford, J.M.; Roberts, P.E. 1978: Distribution of reactive phosphorus and plankton in relation to upwelling and surface circulation around New Zealand. *New Zealand Journal of Marine and Freshwater Research 12*: 1–15.

21 Bradbury, I.R.; Snelgrove, P.V.R. 2001: Contrasting larval transport in demersal fish and benthic invertebrates: the roles of behaviour and advective processes in determining spatial pattern. *Canadian Journal of Fisheries and Aquatic Sciences 58*: 811–823.

22 Ridgway, N.M. 1980: Hydrological conditions and circulation off the west coast of the North Island, New Zealand. *New Zealand Journal of Marine and Freshwater Research 14*: 155–167.

23 Booth, J. 1983: Getting the drift of lobster larvae. *Catch '83 September*: 12.

24 Booth, J.D.; Phillips, B.F. 1994: Early life history of spiny lobster. *Crustaceana 66*: 271–294.

25 Roemmich, D.; Sutton, P. 1998: The mean and variability of ocean circulation past northern New Zealand: determining the representativeness of hydrographic climatologies. *Journal of Geophysical Research 103*: 13041–13054.

26 Williamson, D.I. 1967: Some recent advances and outstanding problems in the study of larval Crustacea. *Marine Biological Association of India Proceedings of the Symposium on Crustacea—Part II*: 815–823.

27 Chiswell, S.M.; Wilkin, J.; Booth, J.D.; Stanton, B. 2003: Trans-Tasman Sea larval transport: is Australia a source for New Zealand rock lobsters? *Marine Ecology Progress Series 247*: 173–182.

28 Brasher, D.J.; Ovenden, J.R.; Booth, J.D.; White, R.W.G. 1992: Genetic subdivision of Australian and New Zealand populations of *Jasus verreauxi* (Decapoda: Palinuridae)—preliminary evidence from the mitochondrial genome. *New Zealand Journal of Marine and Freshwater Research 26*: 53–58.

29 Kennan, S.C.; Pinkerton, M. 2008: Seasonal cycle of ocean colour around New Zealand. *Eos, Transactions of the American Geophysical Union, West Pacific Geophysics Meeting, Abstract OS44A-02.*

30 Jeffs, A.G.; Montgomery, J.C.; Tindle, C.T. 2005: How do spiny lobster post-larvae find the coast? *New Zealand Journal of Marine and Freshwater Research 39*: 605–617.

31 Perry, L. 1999: Sediment effects on the behavior and survival of young juvenile *Jasus edwardsii*. *The Lobster Newsletter 12(1)*: 4–5.

32 Montgomery, S.S.; Craig, J.R. 2005: Distribution and abundance of recruits of the eastern rock lobster (*Jasus verreauxi*) along the coast of New South Wales, Australia. *New Zealand Journal of Marine and Freshwater Research 39*: 619–628.

33 Chiswell, S.M.; Booth, J.D. 1999: Rock lobster *Jasus edwardsii* larval retention by the Wairarapa Eddy off New Zealand. *Marine Ecology Progress Series 183*: 227–240.

34 Dennis, D.M.; Ye, Y.; Pitcher, C.R.; Skewes, T.D. 2004: Ecology and stock assessment of the ornate rock lobster *Panulirus ornatus* population in Torres Strait, Australia. In: *Spiny lobster ecology and exploitation in the South China Sea region.* Proceedings

of a workshop held at the Institute of Oceanography, Nha Trang, Vietnam, July 2004. Williams, K.C. (ed) pp 29–40. CSIRO Australian Centre for International Agricultural Research, Canberra.

35 Sekiguchi, H.; Inoue, N. 2002: Recent advances in larval recruitment processes of scyllarid and palinurid lobsters in Japanese waters. *Journal of Oceanography 58*: 747–757.

36 Phillips, B.F.; Chubb, C.F.; Melville-Smith, R. 2000: The status of Australia's rock lobster fisheries. *In*: *Spiny lobsters: fisheries and culture*. Phillips, B.F; Kittaka, J. (eds) pp 45–77. Fishing News Books, Oxford.

37 Groeneveld, J.C.; Goni, R.; Latrouite, D. 2006: *Palinurus* species. *In*: *Lobsters: biology, management, aquaculture and fisheries*. Phillips, B.F. (ed) pp 385–411. Blackwell Publishing Ltd, Oxford.

38 Roydhouse, N. 1988: Diver's ear pain or claws two. *SPUMS Journal 18(1)*: 32–33.

39 Booth, J. 1991: Intertidal settlement by pueruli of *Jasus edwardsii* in New Zealand. *The Lobster Newsletter 4(2)*: 1, 3.

5. A CLOSER LOOK AT PACKHORSE

1 Booth, J.D. 2006: *Jasus* species. *In*: *Lobsters: biology, management, aquaculture and fisheries*. Phillips, B.F. (ed) pp 340–358. Blackwell Publishing Ltd, Oxford.

2 Paterson, N.F. 1968: The anatomy of the Cape rock lobster, *Jasus lalandii* (H. Milne Edwards). *Annals of the South African Museum 51*.

3 Auerswald, L.; Gäde, G. 2005: The West Coast rock lobster *Jasus lalandii* as a valuable source for chitin and astaxanthin. *African Journal of Marine Science 27*: 257–264.

4 Ceccaldi, H.J. 2006: The digestive tract: anatomy, physiology, and biochemistry. *In*: *Treatise on Zoology—anatomy, taxonomy, biology. The Crustacea. Volume 2*. Forest, J.; von Vaupel Klein, J.C. (eds) pp 85–203. Brill, Leiden.

5 Mikami, S.; Takashima, F. 2000: Functional morphology of the digestive system. *In*: *Spiny lobsters: fisheries and culture*. Phillips, B.F.; Kittaka, J. (eds) pp 601–610. Fishing News Books, Oxford.

6 Mayrat, A.; McMahon, B.R.; Tanaka, K. 2006: The circulatory system. *In*: *Treatise on Zoology—anatomy, taxonomy, biology. The Crustacea. Volume 2*. Forest, J.; von Vaupel Klein, J.C. (eds) pp 3–84. Brill, Leiden.

7 Meyer-Rochow, V.B.; Tiang, K.M. 1984: The eye of *Jasus edwardsii* (Crustacea, Decapoda, Palinuridae): electrophysiology, histology, and behaviour. *Zoologica 45*: 1–58.

8   Talbot, P.; Helluy 1995: Reproduction and embryonic development. *In*: *Biology of the lobster Homarus americanus*. Factor, J.R. (ed) pp 177–216. Academic Press, New York.

9   Juanes, F.; Smith, L.D. 1995: The ecological consequences of limb damage and loss in decapod crustaceans: a review and prospectus. *Journal of Experimental Marine Biology and Ecology 193*: 197–223.

10 Melville-Smith, R.; de Lestang, S. 2007: Changes in egg production of the western rock lobster (*Panulirus cygnus*) associated with appendage damage. *Fisheries Bulletin 105*: 418–425.

11 Dawkins, R. 1986: *The blind watchmaker: why the evidence of evolution reveals a universe without design.* Norton and Company, Inc.

12 Childress, M.J.; Jury, S.H. 2006: Behaviour. *In*: *Lobsters: biology, management, aquaculture and fisheries*. Phillips, B.F. (ed) pp 78–112. Blackwell Publishing Ltd, Oxford.

13 Butler, M.J.; Steneck, R.S.; Herrnkind, W.F. 2006: Juvenile and adult ecology. *In*: *Lobsters: biology, management, aquaculture and fisheries*. Phillips, B.F. (ed) pp 263–309. Blackwell Publishing Ltd, Oxford.

14 Pollock, D.E. 1997: Egg production and life-history strategies in some clawed and spiny lobster populations. *Bulletin of Marine Science 61*: 97–109.

15 Bradford, J.M.; Roberts, P.E. 1978: Distribution of reactive phosphorus and plankton in relation to upwelling and surface circulation around New Zealand. *New Zealand Journal of Marine and Freshwater Research 12*: 1–15.

16 Baisre, J.A. 1994: Phyllosoma larvae and the phylogeny of Palinuroidea (Crustacea: Decapoda): a review. *Australian Journal of Marine and Freshwater Research 45*: 925–944.

6. DAILY ROUTINES

1   MacDiarmid, A.B. 2007: Personal communication. (National Institute of Water and Atmospheric Research Ltd, Wellington)

2   Davey, A. 2007: Personal communication. (Russell, Bay of Islands)

3   Sykes, D.R. 2007: Personal communication. (New Zealand Rock Lobster Industry Council, Wellington)

4   Montgomery, S.S.; Kittaka, J. 1994: Occurrence of pueruli of *Jasus verreauxi* (H. Milne Edwards, 1851) (Decapoda, Palinuridae) in waters off Cronulla, New South Wales, Australia. *Crustaceana 67*: 65–70.

5   Butler, M.J.; Steneck, R.S.; Herrnkind, W.F. 2006: Juvenile and adult ecology. *In*: *Lobsters: biology, management, aquaculture and fisheries.* Phillips, B.F. (ed) pp 263–309. Blackwell Publishing Ltd, Oxford.

6   Herrnkind, W.F.; Butler, M.J., IV 1994: Settlement of spiny lobster, *Panulirus argus* (Latreille, 1804), in Florida: pattern without predictability? *Crustaceana 67*: 46–64.

7   Hothuis, L.B.; Sivertsen, E. 1967: The Crustacea Decapoda, Mysidacea and Cirripedia of the Tristan da Cunha Archipelago with a revision of the 'frontalis' subgroup of the genus *Jasus. Results of the Norwegian Scientific Expedition to Tristan da Cunha 1937–1938.* No. 52.

8   Childress, M.J.; Jury, S.H. 2006: Behaviour. *In*: *Lobsters: biology, management, aquaculture and fisheries.* Phillips, B.F. (ed) pp 78–112. Blackwell Publishing Ltd, Oxford.

9   Lavalli, K.; Spanier, E.; Grasso, F. 2007: Behavior and sensory biology of slipper lobsters. *In*: *The biology and fisheries of the slipper lobster.* Lavalli, K.L.; Spanier, E. (eds) pp 133–181. CRC Press, London.

10 Shears, N.I.; Babcock, R.I. 2002: Marine reserves demonstrate top-down control of community structure on temperate reefs. *Oecologia 132*: 131–142.

11 Montgomery, S.S.; Liggins, G.W.; Craig, J.R.; McLeod, J.R. 2009: Growth of the spiny lobster *Jasus verreauxi* (Decapoda: Palinuridae) off the east coast of Australia. *New Zealand Journal of Marine and Freshwater Research 43*: 113–123.

12 Wahle, R.A.; Fogarty, M.J. 2006: Growth and development: understanding and modelling growth variability in lobsters. *In*: *Lobsters: biology, management, aquaculture and fisheries.* Phillips, B.F. (ed) pp 1–44. Blackwell Publishing Ltd, Oxford.

13 Booth, J.D. 1984: Size at onset of breeding in female *Jasus verreauxi* (Decapoda: Palinuridae) in New Zealand. *New Zealand Journal of Marine and Freshwater Research 18*: 159–169.

14 MacDiarmid, A.B.; Sainte-Marie, B. 2006: Reproduction. *In*: *Lobsters: biology, management, aquaculture and fisheries.* Phillips, B.F. (ed) pp 45–77. Blackwell Publishing Ltd, Oxford.

15 Melville-Smith, R.; de Lestang, S. 2005: Visual assessment of the reproductive condition of western rock lobsters (*Panulirus cygnus*). *New Zealand Journal of Marine and Freshwater Research 39*: 557–562.

16 Annala, J.H.; McKoy, J.L.; Booth, J.D.; Pike, R.B. 1980: Size at the onset of sexual maturity in female *Jasus edwardsii* (Decapoda: Palinuridae) in New Zealand. *New Zealand Journal of Marine and Freshwater Research 14*: 217-227.

17 Cockcroft, A.C.; Goosen, P.C. 1995: Shrinkage at moulting in the rock lobster *Jasus lalandii* and associated changes in reproductive parameters. *South African Journal of Marine Science 16*: 195–203.

18 MacDiarmid, A.; Booth, J. 2003: Crayfish. *In*: *The living reef. The ecology of New Zealand's rocky reefs.* Andrew, N.; Francis, M. (eds) pp 120–127. Craig Potton Publishing, Nelson.

19 Pollock, D.E. 1997: Egg production and life-history strategies in some clawed and spiny lobster populations. *Bulletin of Marine Science 61*: 97–109.

20 McKoy, J.L.; Leachman, A. 1982: Aggregations of ovigerous female rock lobsters, *Jasus edwardsii* (Decapoda: Palinuridae). *New Zealand Journal of Marine and Freshwater Research 16*: 141–146.

21 MacDiarmid, A.B.; Hickey, B.; Maller, R.A. 1991: Daily movement patterns of the spiny lobster *Jasus edwardsii* (Hutton) on a shallow reef in northern New Zealand. *Journal of Experimental Marine Biology and Ecology 147*: 185–205.

22 Kelly, S. 2008: There's no place like home. *The Lobster Newsletter 21(2)*: 10–12.

23 Raethke, N. 2001: Rock lobsters: chemosensory communicators? *Aquaculture Update 29, Winter* (National Institute of Water and Atmospheric Research Ltd).

24 Bouwma, P.E.; Herrnkind, W.F. 2009: Sound production in Caribbean spiny lobster *Panulirus argus* and its role in escape during predatory attack by *Octopus briareus*. *New Zealand Journal of Marine and Freshwater Research 43*: 3–13.

25 Shields, J.D.; Stephens, F.J.; Jones, B. 2006: Pathogens, parasites and other symbionts. *In*: *Lobsters: biology, management, aquaculture and fisheries.* Phillips, B.F. (ed) pp 146–204. Blackwell Publishing Ltd, Oxford.

7. THE PAGEANT OF FISHING

1 Almog-Shtayer, G.; Spanier, E. 1990: Mediterranean lobsters in ancient times. *The Lobster Newsletter 3(2)*: 5, 12.

2 Hopkins, W. 2007: Personal communication. (Houhora, Northland)

3 Eccles, G. 2007: Personal communication. (Kaitaia, Northland)

4 Leach, B.F.; Anderson, A.J. 1979: Prehistoric exploitation of crayfish in New Zealand. *In*: *Birds of a feather. Osteological and archaeological papers from the South Pacific in honour of R.J. Scarlett.* Anderson, A. (ed) British Archaeological Reports, International Series 62: 141–164.

5 Leach, F. 2006: Fishing in pre-European New Zealand. *Archaeofauna* 15. University of Otago.

6 Long, D.S. 1985: Early crayfishing. *Fishing Industry Board Bulletin 84.*

7 Best, E. 1929: Fishing methods and devices of the Maori. *Dominion Museum Bulletin 12.*

8 Hiroa, Te R. 1926: The Maori craft of netting. *Transactions and Proceedings of the New Zealand Institute 56*: 597–646.

9 Samuels, D. 2007: Personal communication. (Wellington)
10 Pauly, D. 1995: Anecdotes and the shifting baseline syndrome of fisheries. *Trends in Ecology and Evolution 10*: 430.
11 Report of the 1937–38 New Zealand Sea Fisheries Investigation Committee. Government Printer, Wellington.
12 Notes of Walter Skrzynksi, Marine Department, Wellington.
13 Anderson, J.; Anderson, M. 2005: White Island – adventure above and below. http://www.mysteriousnewzealand.co.nz
14 Ministry of Agriculture and Fisheries 1978: Regional fishery background papers ahead of introduction of controlled rock lobster fisheries. Wellington (unpublished).
15 Kensler, C.B.; Skrzynksi, W. 1970: Commercial landings of the spiny lobster *Jasus verreauxi* in New Zealand (Crustacea: Decapoda: Palinuridae). *New Zealand Journal of Marine and Freshwater Research 4*: 46–54.
16 Pike, R.B. 1969: Notes on the New Zealand crayfishing industry. Zoology Honours teaching notes, Victoria University of Wellington.
17 Annala, J.H.; King, M.R. 1983: The 1963–73 New Zealand rock lobster landings by statistical area. Fisheries Research Division Occasional Publication: Data Series 11.
18 Ministry of Fisheries Science Group (Comps.) 2009: Report from the Mid-Year Fishery Assessment Plenary, November 2009: stock assessments and yield estimates. New Zealand Ministry of Fisheries.
19 Westley, S. 2008: Personal communication. (Jervis Bay, New South Wales)
20 Food and Agriculture Organization 2007: Capture production 2005. *FAO Yearbook of Fishery Statistics 100/1*.
21 EOLSS 2009: Crabs and lobsters. United Nations Educational, Scientific and Cultural Organization – Encyclopedia of Life Support Systems Theme 5.5. unescoeolss@eolssonline.net
22 Phillips, B.F.; Melville-Smith, R. 2006: *Panulirus* species. *In*: Lobsters: biology, management, aquaculture and fisheries. Phillips, B.F. (ed) pp 359–384. Blackwell Publishing Ltd, Oxford.
23 Booth, J.D. 2006: *Jasus* species. *In*: Lobsters: biology, management, aquaculture and fisheries. Phillips, B.F. (ed) pp 340–358. Blackwell Publishing Ltd, Oxford.
24 Groeneveld, J.C.; Goñi, R.; Latrouite, D. 2006: *Palinurus* species. *In*: Lobsters: biology, management, aquaculture and fisheries. Phillips, B.F. (ed) pp 385–411. Blackwell Publishing Ltd, Oxford.
25 Ceccaldi, H.J.; Latrouite, D. 2000: The French fisheries for the European spiny lobster *Palinurus elephas*. *In*: Spiny lobsters: fisheries and culture. Phillips, B.F.; Kittaka, J. (eds) pp 200–209. Fishing News Books, Oxford.

26 Cruz, R.; Phillips, B.F. 2000: The artificial shelters (*pesqueros*) used for the spiny lobster (*Panulirus argus*) fisheries in Cuba. *In*: *Spiny lobsters: fisheries and culture*. Phillips, B.F.; Kittaka, J. (eds) pp 400–419. Fishing News Books, Oxford.

27 Briones-Fourzán, P.; Lozano-Álvarez, E.; Eggleston, D.B. 2000: The use of artificial shelters (*casitas*) in research and harvesting of Caribbean spiny lobsters in Mexico. *In*: *Spiny lobsters: fisheries and culture*. Phillips, B.F.; Kittaka, J. (eds) pp 420–446. Fishing News Books, Oxford.

28 Melville-Smith, R.; Phillips, B.F.; Penn, J. 2000: Recreational spiny lobster fisheries— research and management. *In*: *Spiny lobsters: fisheries and culture*. Phillips, B.F; Kittaka, J. (eds) pp 447–461. Fishing News Books, Oxford.

29 Coombes, J.W. 1980: Letter to author accompanying photograph of four packhorse.

30 NZPA 2003: Monster crayfish could be 100. *Dominion Post*, 27 September, p A9.

31 Sharp, W.C.; Bertelsen, R.D.; Leeworthy, V.R. 2005: Long-term trends in the recreational lobster fishery of Florida, United States: landings, effort, and implications for management. *New Zealand Journal of Marine and Freshwater Research 39*: 733–747.

32 TAC Committee 2008: Rock lobster fishery Total Allowable Catch Committee report and determination for 2008/09. NSW Department of Primary Industries.

8. MANAGING THE FORTUNES OF FISHERIES

1 Lipcius, R.N.; Eggleston, D.B. 2000: Ecology and fishery biology of spiny lobsters. *In*: *Spiny lobsters: fisheries and culture*. Phillips, B.F.; Kittaka, J. (eds) pp 1–41. Fishing News Books, Oxford.

2 Breen, P.A.; Kendrick, T.H. 1997: A fisheries management success story: the Gisborne, New Zealand, fishery for red rock lobsters (*Jasus edwardsii*). *Marine and Freshwater Research 48*: 1103–1110.

3 Haist, V.; Breen, P.A.; Kim, S.W.; Starr, P.J. 2005: Stock assessment of red rock lobsters (*Jasus edwardsii*) in CRA 3 in 2004. New Zealand Fisheries Assessment Report 2005/38.

4 Booth, J.D.; McKenzie, A. 2009: Strong relationships between levels of puerulus settlement and recruited stock abundance in the red rock lobster (*Jasus edwardsii*) in New Zealand. *Fisheries Research 95*: 161–168.

5 Grobler, C.A.F.; Noli-Peard, K.R. 1997: *Jasus lalandii* in post-independence Namibia: monitoring population trends and stock recovery in relation to a variable environment. *Marine and Freshwater Research 48*: 1015–1022.

6 Pollock, D.E.; Cockcroft, A.C.; Groeneveld, J.C.; Shoeman, D.S. 2000: The commercial fisheries for *Jasus* and *Palinurus* species in the south-east Atlantic and

south-west Indian Oceans. *In*: *Spiny lobsters: fisheries and culture*. Phillips, B.F.; Kittaka. J. (eds) pp 105–120. Fishing News Books, Oxford.

7 Pearce, A.F.; Phillips, B.F. 1988: ENSO events, the Leeuwin Current, and larval recruitment of the western rock lobster. *Journal du Conseil International pour l'Exploration de la Mer 45*: 13–21.

8 Phillips, B.F.; Pearce, A.F.; Litchfield, R.; Guzman del Proo, S. 2000: Spiny lobster catches and the ocean environment. *In*: *Spiny lobsters: fisheries and culture*. Phillips, B.F.; Kittaka, J. (eds) pp 321–333. Fishing News Books, Oxford.

9 Clarke, A.J. 2008: *An introduction to the dynamics of El Niño and the Southern Oscillation*. Elsevier, Amsterdam.

10 Polovina, J.J.; Mitchum, G.T. 1992: Variability in spiny lobster *Panulirus marginatus* recruitment and sea level in the Northwestern Hawaiian Islands. *Fishery Bulletin 90*: 483–493.

11 Booth, J.D.; Bradford, E.; Renwick, J. 2000: *Jasus edwardsii* puerulus settlement levels examined in relation to the ocean environment and to subsequent juvenile and recruit abundance. New Zealand Fisheries Assessment Report 2000/34.

12 Phillips, B.F.; Melville-Smith, R. 2006: *Panulirus* species. *In*: *Lobsters: biology, management, aquaculture and fisheries*. Phillips, B.F. (ed) pp 359–384. Blackwell Publishing Ltd, Oxford.

13 Inoue, N.; Sekiguchi, H. 2009: Can long-term variation in catch of Japanese spiny lobster *Panulirus japonicus* be explained by larval supply through the Kuroshio Current? *New Zealand Journal of Marine and Freshwater Research 43*: 89–99.

14 Trenberth, K.E.; Hurrell, J.W. 1994: Decadal atmosphere-ocean variations in the Pacific. *Climate Dynamics 9*: 303–319.

15 TAC Committee 2008: Rock lobster fishery. Total Allowable Catch Committee report and determination for 2008/09. NSW Department of Primary Industries.

16 Melville-Smith, R.; de Lestang, S.; Thomson, A.W. 2009: Spatial and temporal changes in egg production in the western rock lobster (*Panulirus cygnus*) fishery. *New Zealand Journal of Marine and Freshwater Research 43*: 151–161.

17 Chubb, C.F. 1994: Reproductive biology: issues for management. *In*: *Spiny lobster management*. Phillips, B.F.; Cobb, J.S.; Kittaka, J. (eds) pp 181–212. Fishing News Books, Oxford.

18 Smith, M.T.; Addison, J.T. 2003: Methods for stock assessment of crustacean fisheries. *Fisheries Research 65*: 231–256.

19 EOLSS 2009: Crabs and lobsters. United Nations Educational, Scientific and Cultural Organization – Encyclopedia of Life Support Systems Theme 5.5. unescoeolss@eolssonline.net

20 Phillips, B.F.; Booth, J.D. 1994: Design, use, and effectiveness of collectors for catching the puerulus stage of spiny lobsters. *Reviews in Fisheries Science 2*: 255–289.
21 Phillips, B.F.; Cruz, R.; Caputi, N.; Brown, R.S. 2000: Predicting the catch of spiny lobster fisheries. *In: Spiny lobsters: fisheries and culture*. Phillips, B.F; Kittaka, J. (eds) pp 357–375. Fishing News Books, Oxford.
22 Gardner, C.; Frusher, S.D.; Kennedy, R.B.; Cawthorn, A. 2001: Relationship between settlement of southern rock lobster, *Jasus edwardsii*, and recruitment to the fishery in Tasmania, Australia. *Marine and Freshwater Research 52*: 1271–1275.
23 Haist, V.; Breen, P.A.; Starr, P.J. 2009: A multi-stock, length-based assessment model for New Zealand rock lobster (*Jasus edwardsii*). *New Zealand Journal of Marine and Freshwater Research 43*: 355–371.
24 Fishery status reports *In: Spiny lobsters: fisheries and culture*. Phillips, B.F.; Kittaka, J. (eds). Fishing News Books, Oxford.
25 Fonteles-Filho, A.A. 2000: The state of the lobster fishery in north-east Brazil. *In: Spiny lobsters: fisheries and culture*. Phillips, B.F.; Kittaka, J. (eds) pp 121–134. Fishing News Books, Oxford.
26 Breen, P.A. 1997: Ten years of progress in the rock lobster fishery. *Seafood New Zealand November*: 29–31.
27 Breen, P.A.; Kim, S.W.; Starr, P.J.; Bentley, N. 2002: Assessment of the red rock lobsters (*Jasus edwardsii*) in area CRA 3 in 2001. New Zealand Fisheries Assessment Report 2002/27.
28 Starr, P.J.; Breen, P.A.; Kendrick, T.H.; Haist, V. 2009: Model and data used for the 2008 stock assessment of rock lobsters (*Jasus edwardsii*) in CRA 3. New Zealand Fisheries Assessment Report 2009/22.

9. IMPACTS OF FISHING

1 Warne, K. 2008: No-take zone. *New Zealand Geographic 90, March–April*: 58–78.
2 Kelly, S.; Scott, D.; MacDiarmid, A.B.; Babcock, R.C. 2000: Spiny lobster, *Jasus edwardsii*, recovery in New Zealand marine reserves. *Biological Conservation 92*: 359–369.
3 Butler, M.J.; Steneck, R.S.; Herrnkind, W.F. 2006: Juvenile and adult ecology. *In: Lobsters: biology, management, aquaculture and fisheries*. Phillips, B.F. (ed) pp 263–309. Blackwell Publishing Ltd, Oxford.
4 Freeman, D.J. 2008: The ecology of spiny lobsters (*Jasus edwardsii*) on fished and unfished reefs. Unpublished PhD thesis, University of Auckland, New Zealand.
5 Kelly, S.; Scott, D.; MacDiarmid, A.B. 2002: The value of a spillover fishery for spiny

lobsters around a marine reserve in northern New Zealand. *Coastal Management 30*: 153–166.

6   Department of Conservation. New Zealand's marine reserves: protecting our seas tiakina a Tangaroa. www.doc.govt.nz (no date given).

7   Pauly, D.; Watson, R.; Alder, J. 2005: Global trends in world fisheries: impacts on marine ecosystems and food security. *Philosophical Transactions of the Royal Society B 360*: 5–12.

8   Frusher, S.; Gibson, I.J. 1999: Bycatch in the Tasmanian rock lobster fishery. *In*: *Establishing meaningful targets for bycatch reduction in Australian fisheries.* Buxton, C.; Eayrs, S. (eds). Australian Society for Fish Biology, Sydney.

9   Davey, N. 2007: Personal communication. (Russell, Bay of Islands)

10 Royal Forest and Bird Protection Society 2008: *Best fish guide 07–08.*

11 Campbell, R.; Coyne, M.; Caputi, N. 2007: Research and management response to interactions, and accidental mortality, of the Australian sea lion, *Neophoca cinerea*, in the west coast rock lobster fishery. *The Lobster Newsletter 20(1)*: 11–14.

12 Lewis, C.F.; Slade, S.L.; Maxwell, K.E.; Matthews, T.R. 2009: Lobster trap impact on coral reefs: effects of wind-driven trap movement. *New Zealand Journal of Marine and Freshwater Research 43*: 271–282.

13 Eno, N.C. and 7 others 2001: Effects of crustacean traps on benthic fauna. *ICES Journal of Marine Science 58*: 11–20.

14 Casement, D.; Svane, I. 1999: *Direct effects of rock lobster pots on temperate shallow rocky reefs in South Australia.* South Australian Research & Development Institute.

15 Polovina, J.J. 2000: The lobster fishery in the North-western Hawaiian Islands. *In*: *Spiny lobsters: fisheries and culture.* Phillips, B.F.; Kittaka, J. (eds) pp 98–104. Fishing News Books, Oxford.

16 Freeman, D.J.; MacDiarmid, A.B. 2009: Healthier lobsters in a marine reserve: effects of fishing on disease incidence in the spiny lobster, *Jasus edwardsii*. *Marine and Freshwater Research 60*: 140–145.

17 EOLSS 2009: Crabs and lobsters. United Nations Educational, Scientific and Cultural Organization – Encyclopedia of Life Support Systems Theme 5.5. unescoeolss@eolssonline.net

18 Juanes, F.; Smith, L.D. 1995: The ecological consequences of limb damage and loss in decapod crustaceans: a review and prospectus. *Journal of Experimental Marine Biology and Ecology 193*: 197–223.

19 Meyer-Rochow, V.B. 1994: Light-induced damage to photoreceptors of spiny lobsters and other crustaceans. *Crustaceana 67*: 95–109.

20 Morgan, G. 2008: Carbon management in lobster fisheries. *The Lobster Newsletter 21(2)*: 6–10.

21 Doney, S.C.; Fabry, V.J.; Feely, R.A.; Kleypas, J.A. 2009: Ocean acidification: the other CO_2 problem. *Annual Review of Marine Science 1*: 169–192.

10. GREEN FINGERS

1. Kittaka, J. 2000: Culture of larval spiny lobsters. *In*: *Spiny Lobsters: fisheries and culture*. Phillips, B.F; Kittaka, J. (eds) pp 508–532. Fishing News Books, Oxford.
2. Kittaka, J. 2007: Personal communication. (Chiba, Japan)
3. Moss, G., James, P.; Tong, L. 2000: *Jasus verreauxi* phyllosomas cultured. *The Lobster Newsletter 13(1)*: 9–10.
4. Moss, G.A.; James, P.J.; Allen, S.E.; Bruce, M.P. 2004: Temperature effects on the embryo development and hatching of the spiny lobster *Sagmariasus verreauxi*. *New Zealand Journal of Marine and Freshwater Research 38*: 795–801.
5. Moss, G.A.; Tong, L.J.; Allen, S.E. 2001: Effect of temperature and food ration on the growth and survival of early and mid-stage phyllosomas of the spiny lobster *Sagmariasus verreauxi*. *Marine and Freshwater Research 52*: 1459–1464.
6. Ritar, A. 2008: Personal communication. (University of Tasmania, Hobart)
7. Jeffs, A. 2008: Personal communication. (University of Auckland)
8. Booth, J.D.; Kittaka, J. 2000: Spiny lobster growout. *In*: *Spiny lobsters: fisheries and culture*. Phillips, B.F; Kittaka, J. (eds) pp 556–585. Fishing News Books, Oxford.
9. Bell, J.D.; Rothlisberg, P.C.; Munro, J.L.; Loneragan, N.R.; Nash, W.J.; Ward, R.D.; Andrew, N.L. 2005: Restocking and stock enhancement of marine invertebrate fisheries. *Advances in Marine Biology 49*: 1–358.
10. Oliver, M.D. 2008: Personal communication. (NIWA, Wellington)
11. Oliver, M.D.; MacDiarmid, A.B.; Stewart, R.A.; Gardner, C. 2008: Anti-predator behavior of captive-reared and wild juvenile spiny lobster (*Jasus edwardsii*). *Reviews in Fisheries Science 16*: 186–194.
12. Gardner, C.; van Putten, E.I. 2008: The economic feasibility of translocating rock lobsters to increase yield. *Reviews in Fisheries Science 16*: 154–163.
13. Butler, M.J. IV; Herrnkind, W.F. 1997: A test of recruitment limitation and the potential for artificial enhancement of spiny lobster (*Panulirus argus*) populations in Florida. *Canadian Journal of Fisheries and Aquatic Sciences 54*: 452–63.
14. Nonaka, M.; Fushimi, H.; Yamakawa, T. 2000: The spiny lobster fishery in Japan and restocking. *In*: *Spiny lobsters: fisheries and culture*. Phillips, B.F; Kittaka, J. (eds) pp 221–242. Fishing News Books, Oxford.
15. Shizuoka Prefecture 1979: Man-made nursery for Japanese spiny lobsters in South Izu. Brochure in Japanese.

11. THE FINAL CHAPTER

1. Stephens, F.; Fotedar, S.; Evans, L. 2003: *Rock lobster health and diseases: a guide for the lobster industry.* Aquatic Science Research Unit Curtin University of Technology.
2. Crear, B.; Cobcroft, J.; Battaglene, S. 2003: *Recirculating systems for holding rock lobsters. Guide for the rock lobster industry 2.* Tasmanian Aquaculture and Fisheries Institute, University of Tasmania.
3. Stevens, R.N.; Sykes, D. 2000: Export marketing of Australian and New Zealand spiny lobsters. *In*: *Spiny lobsters: fisheries and culture.* Phillips, B.F; Kittaka, J. (eds) pp 641–653. Fishing News Books, Oxford.
4. Konosu, S.; Yamaguchi, K. 2000: Colour and taste. *In*: *Spiny lobsters: fisheries and culture.* Phillips, B.F; Kittaka, J. (eds) pp 625–632. Fishing News Books, Oxford.
5. Chandrapavan, A.; Gardner, C.; Linnane, A.; Hobday, D. 2009: Colour variation in the southern rock lobster *Jasus edwardsii* and its economic impact on the commercial industry. *New Zealand Journal of Marine and Freshwater Research 43*: 537–545.
6. Tsuruta, M.; Kittaka, J. 2000: Marketing and distribution in Japan. *In*: *Spiny lobsters: fisheries and culture.* Phillips, B.F; Kittaka, J. (eds) pp 654–663. Fishing News Books, Oxford.
7. Royal New Zealand Society for the Prevention of Cruelty to Animals 2004: *Animal welfare policy. Crayfish and rock lobster.* www.rnzspca.org.nz
8. Lowe, T.E.; Gregory, N.G. 1999: A humane end for lobsters. *NZ Science Monthly September*: 11.

EPILOGUE

1. Royal Forest and Bird Protection Society 2010: *Best Fish Guide 2009–10.*

SELECTED GLOSSARY OF TECHNICAL TERMS

Abundance index: a quantitative measure of lobster density or abundance, usually presented as a time series.
Autotomy: the reflex severance of an appendage from the body, especially when the organism is injured or under attack.
B_{MSY}: the average biomass of lobsters that results from taking an average catch of maximum sustainable yield under various types of harvest strategy.
Bathymetry: the measurement of the depth of ocean or lake floors.
Benthic: relating to the bottom of a sea or lake, or the organisms that live there.
Berried: bearing external eggs (ovigerous).
Biomass: the size of the lobster stock in units of weight. Often biomass refers to only one part of the stock: spawning biomass, recruited biomass, etc.
Bog: *see* spermatophore.
Catch per unit effort (CPUE): the quantity of lobsters caught with one standard unit of fishing effort (usually kilograms per pot lift).
Catch sample: a sample taken from a (usually commercial) catch at sea.
Chitin: the long-chain polymer of a derivative of glucose found widely in the natural world, including the exoskeleton of many crustaceans.
Cohort: a group of lobsters of the same stock that settled in the same year. Also known as a 'year class'.
Contranatant migration: the migration of juvenile lobsters back to the breeding area after being carried away as phyllosomas by currents.
Cuticle: in arthropods, the multi-layered outer skeleton. There are two layers: the epicuticle, which is waxy, without chitin, and water resistant; and the procuticle, composed mainly of chitin. The procuticle is much thicker than the epicuticle, with two layers, the outer exocuticle and the inner endocuticle.
Deep-scattering layer: a horizontal zone of living organisms, including plankton and fishes, occurring beneath the surface of many ocean areas, so-called because the layer scatters or reflects sound waves, causing echoes in depth sounders.
Demographic bottleneck: in this context, the shortage of a key ecological prerequisite, such as shelters of the appropriate size, which significantly reduces the levels of survival of a life-history phase.
El Niño: a prolonged warming by at least 0.5°C of the surface temperature of the east-central tropical Pacific Ocean. It occurs on average every 5 years and lasts 9–24 months. (cf., La Niña, a prolonged cooling.)

El Niño and the Southern Oscillation (ENSO): a climate pattern occurring across the tropical Pacific Ocean about every 5 years and composed of both an oceanic component (El Niño and La Niña, according to phase) and an atmospheric one characterised by prolonged changes in surface pressure across the tropical Pacific (Southern Oscillation). Often abbreviated to El Niño. *See* El Niño; Southern Oscillation Index.

Endopod: the inner branch (ramus) of a biramous appendage, especially one arising from the basis or from the protopodite of the pleopod.

Exopod: the outer or lateral branch (ramus) of a biramous appendage, especially one arising from the basis or from the protopodite of the pleopod.

Fecundity: the number of eggs in a brood (clutch).

FV: fishing vessel.

Growth overfishing: when lobsters are harvested at an average size smaller than the size that would produce the maximum yield in weight per recruit.

Holotype: the type specimen—the original specimen(s) on which the description and naming of a new species is based. If unknown or lost, a new one is selected, the lectotype; if all original material is lost or destroyed, the neotype.

Indo-West Pacific: a zoological region normally taken to span the entire Indian Ocean including the Red Sea, and the Pacific Ocean as far east as the Caroline Islands but short of the Marshall Islands.

ITQ: individual transferable quota.

Larva: a distinct juvenile form many animals undergo before metamorphosis into the postlarval form, or into the adult. It usually looks very different from the adult, and often occupies a different habitat.

Length composition or length frequency: the range of lobster sizes (by length) in a catch or market sample.

Market sample: a sample taken from a (usually commercial) landed catch.

Maximum sustainable yield (MSY): the largest long-term average catch or yield that can be taken from a lobster stock, under prevailing ecological and environmental conditions, without impairing its renewability through natural growth and reproduction.

Metamorphosis: a period of rapid and dramatic transformation in form; in spiny lobsters the moult from the final phyllosoma instar to the puerulus.

MLS: minimum legal size.

MPA: marine protected area.

MV: motor vessel.

MYA: million years ago.

Naupliosoma: the first form of the lobster larva after hatching. It seems to be present in only the more primitive lobster species.

Nekton: the aggregate of actively swimming organisms in a body of water able to move independently of water currents.

New recruit: an individual lobster that recently entered a particular (usually the fished) component of a stock.

Phyllosoma: the long-lived, leaf-like, transparent, planktonic larval form of spiny lobsters and slipper lobsters.

Plankton: the aggregate of drifting organisms that inhabit the pelagic zone of a body of water. The plant component is the phytoplankton; the animals are the zooplankton.

Population: a geographically distinct grouping of individual lobsters of a particular species.

Postlarva: a phase following the larval phase, usually similar in form to the adult, and characterised by the use of abdominal appendages (pleopods) for propulsion; in spiny lobsters, the puerulus.

Pre-recruit: an individual that has not yet entered the harvestable component of a lobster stock because it is still too small. The term is normally applied to those only a little undersized.

Projection: a prediction about trends in stock size and fishery dynamics. Long-term projections are usually uncertain because the results are strongly dependent on recruitment, which is very difficult to predict accurately.

Puerulus: the postlarval phase of spiny lobster development, transitional between the phyllosoma and the juvenile.

QMS: Quota Management System.

Recruit: *see* new recruit.

Recruitment: depending on context, the settlement of pueruli (larval or postlarval recruitment); the addition of new lobsters to the harvestable component of a stock (recruitment to the fishery), or the addition of new lobsters to the breeding population.

Recruitment mechanism: a conceptual model of how new individuals of a species are added to its breeding population.

Recruitment overfishing: when fishing is so heavy that the adult (spawning) population is reduced to the point that it does not have the reproductive capacity to replenish itself.

Rock lobster: *see* spiny lobster.

RV: research vessel.

Scaphognathite: a thin, leaf-like appendage (exopod) of the second maxilla of decapod crustaceans, which draws water through the gill cavity.

Seamount: a submarine mountain that projects at least 250 metres above the seafloor.

SeaWiFS Project: Sea-viewing Wide Field-of-view Sensor Project that provides quantitative data on global ocean bio-optical properties.

Settlement: when the puerulus stops any further extensive forward swimming and takes up residence on the seafloor.

Size at onset of maturity (SOM) and size at onset of breeding (SOB): the mean size at which lobsters in a particular population mature (SOM) or first bear eggs (SOB). Usually, but not always, SOM and SOB are the same.

Size composition: *see* length composition.

Slipper lobster: a decapod crustacean belonging to the family Scyllaridae.

Southern Oscillation Index (SOI): the anomalous monthly surface atmospheric pressure at Tahiti divided by its standard deviation minus the anomalous monthly surface atmospheric pressure at Darwin divided by its standard deviation. A measure of differences in surface atmospheric pressure across the tropical Pacific Ocean.

Spawning: in spiny lobsters, the extrusion of eggs (egg-laying).

Spawning stock: that part of a population which is capable of reproducing.

Spermatophore: in spiny lobsters, a mass (sperm packet) containing spermatozoa transferred in its entirety to the female sternum. Also referred to as a bog, or tarspot.

Spiny lobster: a decapod crustacean belonging to the family Palinuridae.

Stock (biological): the population of a given species forming a reproductive unit that crosses little if at all with other units.

Stock (nominal): the population of a given species forming an assessment or management unit.

Stock assessment: the application of statistical and mathematical tools to fishery data in order to obtain a quantitative understanding of the status of the stock relative to defined benchmarks or reference points such as B_{MSY}.

Stridulating organ: a noise-making structure at the base of the antennae, present in all spiny lobsters except *Jasus*, *Palinurellus*, *Projasus* and *Sagmariasus* (the Silentes). Spiny lobsters with a stridulating organ belong to the Stridentes.

Sustainability: the ability of a lobster stock to persist in the long-term, usually, at the same time, producing near its maximum potential.

TAC: total allowable catch.

TACC: total allowable commercial catch.

Trophic: of or relating to feeding and nutrition.

Trophic level: the position of an organism in the food chain, e.g. plants, herbivores, carnivores.

Year class: *see* cohort.

INDEX

Bold indicates a higher-level entry; italics mean it's a map

Aberdein, Alan **63**
Achelata **47**, **55**
African spear lobster *see Linuparus somniosus*
age and growth 116, 117, **126**, **127**, 142, 181, 196, 204, 206
   Jasus edwardsii 127, 207
   Jasus lalandii 129
   Palinurus elephas 127, 204, 205
   Panulirus argus 127, 136, 204
   Panulirus cygnus 40, 204
   Panulirus homarus 204
   Panulirus interruptus 204
   Panulirus ornatus 204, 205
   Sagmariasus verreauxi 117, **126–128**, 202, 204
aka-ebi *see Panulirus brunneiflagellum*
Albatros **65–67**
American blunthorn lobster *see Palinustus truncatus*
Amsterdam Island *62*, 69
Antarctus mawsoni **75**
antennal gland **112–114**
Antipodarctus aoteanus **75**, **93**
aquaculture 110, **199–206**
   collecting pueruli for ongrowing **205**
   eyestalk ablation **206**
   food conversion ratio **204**, 206
   Jasus edwardsii 199, 201
   Jasus lalandii 199, 200
   Jeffs, Andrew 205
   Kittaka, Jiro **89**, **198–201**
   Palinurus elephas 199, **204**, **205**
   Panulirus argus 204
   Panulirus cygnus 204
   Panulirus homarus 204
   Panulirus interruptus 204
   Panulirus japonicus 199, 202
   Panulirus ornatus 202, 204, **205**
   Ritar, Arthur **202**, 203
   Sagmariasus verreauxi 94, **199–204**
Arabian whip lobster *see Puerulus sewelli*

Arctides antipodarum (Spanish lobster) **75**, **93**
artificial shelter/habitat/reef 158, **161–163**, **206–209**
Asian blunthorn lobster *see Palinustus holthuisi*
assessment model **180–185**
attraction hypothesis **163**
Australia
   Australian sea lion bycatch **194**
   El Niño-Southern Oscillation **168**, **169**
   Haswell, William **49**
   Jasus edwardsii 35, **62**, 97, **158**, *159*, 160, **178**, **197**, 199, 211, **218**
   Kittaka, Jiro 199
   Panulirus cygnus 40, 97, **107**, 116, **128**, **158**, *159*, **165**, **168**, 172, **176**, **178–180**, **186**
   Panulirus ornatus **107**, *159*, 177, 202
   phyllosoma culture 89, **202**, **203**
   recreational fishing **165**
   Sagmariasus verreauxi 34, **51**, **55**, **56**, **60**, *61*, 87, **89**, **103**, *105*, 106, 119, **156**, **157**, **165**, **171**, 172,178, 179, 199, 200, **202**, **203**
   sinking of *Southern Raider* 68
autotomy **116**, **243**

bacterial shell disease **135**, **136**
banded whip lobster *see Puerulus angulatus*
Bay of Islands *26*, 59, 120, 121, 137, 144, 150, *152*
Bay of Plenty *26*, 59, 60, 87, **152–154**
behaviour
   adult **123–125**, **129–131**, **133–135**
   juvenile *104*, *105*, **120–124**, **206**, **207**
   phyllosoma 85, **91**, **94**, **102**, **103**, **118**
   puerulus **94**, 95, *104*, *105*, 106, *107*, **108**, 120, **121**
Best, Elsdon **147**, 148
blue spiny lobster *see Panulirus inflatus*
Bream Bay *26*, 31, **35**, **165**
breeding **127–133**

breeding area 27, **34–36**, 39, 40, 59, **60**, 97, **103**, *104*, *105*, 106, **107**, 118, 133, 220
brood size **87**, **117**, **132**
egg hatching **39**, **40**, 86, 87, **97**, 118, **133**, **173**, **201**
egg laying 39, 40, **87**, **88**, **116**, 129, **130**
egg production **117**, 171, 172, **176**, **189**, **190**, 196
fertilisation 116, 128, **130–132**, 173, **174**
incubation **87**, 116, **132**, **133**
mating 40, 128, **129–131**, 196
reproductive system **114–116**
size at onset of maturity/breeding 37, **117**, **127–129**, 171, 172, 196
spawning/breeding stock **157**, **171–174**, **176**, 177, **189**
Briones-Fourzan, Patricia **163**
brood period *see* incubation
brown spiny lobster *see Panulirus echinatus*
Buchanan, John **48**, 49
Buck, Peter **148**, **149**
buffalo blunthorn lobster *see Palinustus mossambicus*
bycatch 143, **160**, **192–194**, 221

California spiny lobster *see Panulirus interruptus*
cannibalism **126**, 174
Cape Egmont 26, 27, *61*
Cape jagged lobster *see Projasus parkeri*
Cape Maria van Diemen 26, **33** 34, **39**, 59, **96**, 146
Cape Palliser 147
Cape Reinga 25, **26**, 27, **31**, 33, 34, **39**, 61, **96**, **99**, **100**, **105**, 109, 138, 139, **141**, 145, *152*, **155**
Cape rock lobster *see Jasus lalandii*
Cape Rodney-Okakari Point Marine Reserve **134**, **189–192**
Cape Verde spiny lobster *see Palinurus charlestoni*
carbon management **197**
Caribbean spiny lobster *see Panulirus argus*
casitas cubana **161–163**
Castlepoint 26, **106**, **108**, *167*, **184**
catch per unit effort (CPUE) 157, 166, 167, **175**, **176**, 181, **183**, **184**, **243**
catch prediction **168–170**, **177–180**, **183–186**
catch sample **127**, **128**, **154**, 176, **243**
Cavalli Islands 149, **150**, 151, *152*, 153, **154**

Cave, Joe **70**, **71**
Chadderton, John **64**, **65–71**, **73**
Chatham Islands **60**, *61*, 147, **159**, 167
Chilean jagged lobster *see Projasus bahamondei*
Chinese spiny lobster *see Panulirus stimpsoni*
Chiswell, Steve **102**
circulatory system **114**, 213, 216
clawed lobster **44**, **55**, 129, 207
climate change **197**
collector *see* puerulus collector
colour of shell **44**, 85, **110**, **117**, 138, 204, 206, 207, 211, **218**
contranatant migration **34**, *35*, *36*, **37**, 39, 86, *104*, *105*, *107*, **243**
Cook Strait 26, 31, 32, 34, **60**, 61, 102
cooking 139, **149**, 211, **212**, 217, 218, **219**
Coromandel Peninsula 26, 59
Cox, Serena **91**
CRA 3 **166**, *167*, **182–184**
CRA 7 *167*, **181–185**
CRA 8 *167*
crayfish 43, **44**, 48, **55**, **149**, 153, 154, 167
crustacean 38, 44, **54**, 69, **73**, **85**, **123**, **124**, 136, 187, **193**, 202
Cruz, Raoul **161**
Cuba 127, **158**, **159**, **161**, **162**, **178**
culture *see* aquaculture
customary fishing **147–151**, 156, 174, 181

Davey, Adam **121**, 137, **143**
Davey, Nat 27, 137, **194**, **215**
decapod **44**, 45, **47**, **54**, 110, 115, 134, 217
deepwater spiny lobster *see Projasus parkeri*
demographic bottleneck **122**, **206**, **208**, **244**
digestive gland 94, **96**, **112**, **114**, **181**, **203**, **212**
digestive system *also see* nutrition **112**, **114**, 126, 213
disease **135**, **136**, **196**, 201, **202**, 204, 207
distribution **59–83**
  Jasus caveorum **62**, **70**, 71
  Jasus edwardsii 35–37, **62**, **64**, *159*, 160
  Jasus frontalis **62**
  Jasus lalandii **62**, 97, *159*
  Jasus paulensis **62**, **64**, **68**, 69
  Jasus tristani **62**, 63
  Palinurus barbarae **64**, *79*
  Palinurus delagoae **64**, *79*, *107*
  Palinurus elephas **73**, *79*, *159*
  Palinurus gilchristi *79*, *107*, *159*

Panulirus argus **80**, **158**, *159*
Panulirus cygnus **80**, 97, *107*, *159*
Panulirus interruptus **80**, *159*
Panulirus japonicus **81**, *107*
Panulirus marginatus **80**
Projasus bahamondei **64**, 73, *78*, *159*
Projasus parkeri **64**, 69, *78*, *159*
Sagmariasus verreauxi 25–35, **59–61**, 102–105

early benthic juvenile (EBJ) 86, **122**, 123
East Auckland Current 38, *98*, **101**, 103
East Australian Current 103, **105**
East Cape *26*, 31, **33**, **60**, *61*, 98, 101, 121
Easter Island spiny lobster *see Panulirus pascuensis*
eastern rock lobster *see Sagmariasus verreauxi*
El Niño and El Niño-Southern Oscillation (ENSO) *also see* ocean climate 129, **168–170**, 180, 183, **244**
enhancement 199, **206–209**
enhancement hypothesis **163**
escape gap 70, **171**, **193**
European spiny lobster *see Palinurus elephas*
evolution *see* phylogeny and evolution
excretory system **112**, 113
exoskeleton 40, **94**, **110**, 112, **117**, 120, **127**, **135**, **136**, 142, **143**, 218, 243
exporting *also see* marketing *and* packout 138, 146, 153, **155**, 207, **212–218**
eye 55, 87–90, 94, 95, **108**, 110, 111, **115**, **134**, **196**, 206
eyestalk ablation **206**

Far North **25–35**, *26*, 37, **39**, **40**, 59, **96–105**, **117–119**, 133, **137–146**, **153–156**, **161**, **187**, **188**, 190, **193**, **221**
fecundity 87, **117**, **132**, **244**
fertilisation 116, 128, **130–132**, **173**, **174**, 200
Fiordland *26*, 35, *36*, 147
fishery prediction *see* catch prediction
fishing and fisheries
 Jasus caveorum **70**, 71
 Jasus edwardsii 31, **36**, 64, 70, **140**, **147–151**, 153, **158**, *159*, 160, 166, *167*, **171**, **174**, 178, **181–186**, **193**, **197**
 Jasus lalandii **158**, *159*, **166–168**
 Jasus paulensis **65–70**, 123
 Jasus tristani **63**

Palinurus delagoae **64**, *107*, **160**
Palinurus elephas *159*, **160**
Palinurus gilchristi **107**, *159*
Panulirus argus 158, *159*, 161, 162, 164, **182**, **195**, **196**
Panulirus cygnus 40, 116, *158*, *159*, **165**, 169, **178–180**, **186**
Panulirus interruptus **159**
Panulirus japonicus *159*, 160, 162, 208, **209**
Panulirus laevicauda *159*, **182**
Panulirus ornatus 107, *159*, **161**
Projasus bahamondei **64**, *159*
Projasus parkeri **64**, 69, **70**, **74**, *159*
Puerulus sewelli *159*, **162**
Sagmariasus verreauxi 25–31, 33, 39, 40, 59–61, **137–157**, **161**, **163–165**, 171, **174**, 176, 188, **192–194**, **215**, **220–222**
fishing mortality rate (*F*) **176**
flavour *see* taste
Florida 123, 136, 158, 159, 164, 165, 178, **195**, **196**, **208**
food
 juvenile and adult **124–126**, 204, 205
 phyllosoma 91, 94, **200–203**
forecasting catches *see* catch prediction
Foundation Seamount Chain **62**, 70, 71
Foundation Seamounts rock lobster *see Jasus caveorum*
Foveaux Strait *26*, **34**, 36
Freeman, Debbie **126**, **190**, **195**
freshwater crayfish **44**, 55
furry lobster *see Palinurellus gundlachi*

gas bubble disease **215**
George, Ray **55**
ghost fishing **195**, **196**
Gisborne *26*, 33, *35*, 84, **148**, 166, *167*, 168, 182, 183, **184**, 190, 207
Gough Island **62**
Great Barrier Island *26*, 59, **152**, **154**
Great Exhibition Bay 25, *26*, **98–100**
green spiny lobster *see Panulirus gracilis*
growout **204**, 205
growth *see* age and growth
growth overfishing **182**, **244**

Halfmoon Bay *167*, 183
hand gathering **147**, **148**, **154**, **158**, **160**, **161**, **163–165**, **194**, **205**

Haswell, William **49**
Hawaiian spiny lobster *see Panulirus marginatus*
Heard Island *62*, 68, **69**, **73**
Heller, Camil **48**
Hiroa, Te Rangi *see* Peter Buck
holding pot **145**, **213**, **215**
holotype **42**, **47**, **48**, **50**, **54**, **244**
Holthuis, Lipke **42**, 50, **51**, 76
Hopkins, Bill 39, 143, **144–146**, 147, **155**, 176

Ibacus alticrenatus (prawn killer) **75**, **93**
Ibacus brucei **75**
Ikatere 28, 29
incubation 87, 116, **132**, **133**
Indian Ocean 48, 56, **64–66**, 68–70, **73**, 74, 169
individual transferable quota (ITQ) 144, **174**, 191
Indo-Pacific furry lobster *see Palinurellus wieneckii*
input control **171–174**
instar 87, 88–91, 94, 95, **122**, 179, 203, 245

Japan 89, *107*, *159*, **164**, 198, **199–201**, **208**, **209**, 210, **211**, **212**, 218
Japanese blunthorn lobster *see Palinustus waguensis*
Japanese spear lobster *see Linuparus trigonus*
Japanese spiny lobster *see Panulirus japonicus*
Jasus 49–**53**, **56–58**, 61, **62**, 68, 70, 71, **73**, 74, **76**, 92, **93**, 128, **130**, **158**, *159*, 166, 199, 211
Jasus caveorum (Foundation Seamounts rock lobster) 52, *62*, **70**, **71**, 76
Jasus edwardsii (red *or* southern rock lobster) 76
  age and growth 126, **127**, 207
  Australia *62*, **158**, *159*, **160**, 178, **193**, **197**, 199, **211**, **218**
  behaviour **133–135**
  breeding 97, *107*, **129**, **130**, **134**
  catch prediction 178, **179**, **181–186**
  CRA 3 **166**, *167*, 168, **182–184**
  CRA 7 *167*, **181–185**
  distribution 35, 36, *62*, 64, 97, *159*, 160
  El Niño-Southern Oscillation 170
  enhancement **206–208**
  fishing and fishery 31, 64, **140**, 147–**151**, *159*, 160, 166, *167*, 168, 171, 174, **178**, 180–**186**, **193**, 197
  form 72

juvenile 35, *36*, 37, *104*, *105*, 121, 126, 127, 183, **206–208**
Kittaka, Jiro **198–200**
larval recruitment 118, *107*
marine protected area 126, **134**, **190**, **196**
movement/migration 35, *36*, 37, 40, **133**, **134**, 181, **190**
naupliosoma **133**
phyllosoma, and larval culture **86–88**, **106**, *107*, 198, 199
predator **126**
prey **125**
processing and marketing 211, 214, **216–218**
puerulus and puerulus settlement **95**, **106**, *107*, 108, **166–168**, 170, **178**, **179**, **181–186**
stock assessment **181**, **182**
Jasus flemingi **54**, **55**
Jasus frontalis (Juan Fernández rock lobster) *62*, 76
Jasus lalandii (Cape rock lobster) **40**, **41**, 49, 50, **52**, *62*, 76, 97, **112**, **129**, **158**, *159*, **166–168**, **176**, 199, **200**
Jasus paulensis (St. Paul rock lobster) **52**, *62*, **64–70**, **73**, **76**, **123**
Jasus tristani (Tristan rock lobster) **52**, 61, *62*, **63**, 76
Jasus verreauxi **46**, 49, **51**, **201**
Jeffs, Andrew **37**, **205**
Juan Fernández *62*
Juan Fernández rock lobster *see Jasus frontalis*
Justitia **53**, **73**, **76**, **78**, **92**, **93**
Justitia longimanus 76, *78*
juvenile
  culture **200–206**
  Jasus edwardsii **35–37**, *105*, 121, 126, 127, 183, **206–208**
  Palinurus elephas **127**
  Panulirus argus **40**, **123**, **127**, **178**, **208**
  Panulirus cygnus **40**, *107*, 158, **204**
  Panulirus japonicus **208**, **209**, **211**
  Sagmariasus verreauxi 27–**35**, **44**, **60**, 96, 103, *104*, *105*, **106**, 117, **120–122**, **126**, **127**, **154**, **187**, **188**, **200–202**, **203**

Kaipara Harbour *26*, **99**
Kensler, Craig **128**, **154**
Kerguelen Islands *62*, **67–69**
Kermadec Islands **60**, *61*, **72**, 75, 167

keystone predator **125**, **189**
kill (despatch) **212**, **213**, **217–219**
Kirk, Thomas **48**, **49**, 51
Kittaka, Jiro **89**, **94**, **198–201**
koura 43, **148–151**
koura mara **149**, 151

La Niña **170**, 244
landings *see* fishing and fisheries
larva *see* naupliosoma *and* phyllosoma 245
larval culture *see* phyllosoma culture *and* aquaculture
larval recruitment 86, 96–100, **101–103**, *104*, *105*, **106**, *107*, 118, 168, 169, 177–180, **182–186**
leatherback turtle **194**
Leigh Marine Reserve *see* Cape Rodney-Okakari Point Marine Reserve 26
life history **86–90**, **104**–**107**
Linuparus **52**, **53**, 55, 73, 76, **79**, 92
Linuparus somniosus (African spear lobster) 76, *79*
Linuparus sordidus (Oriental spear lobster) 76, *79*
Linuparus trigonus (Japanese spear lobster) 76, *79*
live holding **145**, **146**, **155**, **156**, 172, 173, **214–217**
live marketing *see* marketing
live transport **146**, **155**, **156**, 213, **214**, 216, 217
longlegged spiny lobster *see Panulirus longipes*

MacDiarmid, Alison **121**, **129**
Madagascar Ridge spiny lobster *see Palinurus barbarae*
Mahia Peninsula 26, 33, *35*
management *also see* regulations 126, 144, **163**, **164**, **166–174**, **180–186**, 192, 220–222
Māori fishing **147–151**
Maria Theresa Reef *62*, *63*, *64*
marine protected area (MPA) 126, 134, **188–192**, 195, **196**
marine reserve *see* marine protected area
market sample **176**, 245
marketing *also see* exporting *and* packout 45, 110, 136, 138, 151, **153**, **155**, **156**, 158, 172, 173, **204**, **205**, 208, 211, **213–218**
Matakaoa Point 26, 31, *35*

Matauri Bay 149–**151**
maturation 32–35, 40, 104–*107*, **114**, **116**, **117**, **127–129**, **154**, **171**, **172**, 196, 200, 212, 213
maximum legal size (MaxLS) **157**, **172–174**, 221
maximum sustainable yield (MSY) **180**, 182, **243**, **245**
megalopa **95**
Mercury Bay **152**, **153**, 187
Mercury Islands 26, 31, *35*, **152–154**
metamorphosis 86, 89, 90, 94, **102–105**, 118, 200, 202, **203**, **245**
Mexico 158, *159*, 161, 162, 164
migration *see* movement/migration
Milne Edwards, Henri 45–**47**, 51, 52
minimum legal size (MLS) 27, 31, 33, 40, 153, **156**, **157**, 158, **163**, 167, **171–173**, 176–178, 180–182, 221
Moeraki *167*, 183
mortality **117**, 126, 132, 160, **176**, 195, 196, 207
Moss, Graeme **201**
moult 29, 39, **40**, 53, **87**, 89, **94**–96, 116, 120, 123, **126**, **127**, 129, 130, 135, **142**, **143**, 149, 175, 179, 200, 201, **206**, 216, 217
movement/migration **34**, **37–40**, 86, *107*, 122–124
    Jasus edwardsii **35**, *36*, **37**, **40**, **133**, 181, **190**
    Jasus lalandii **40**, **41**
    Palinurus delagoae **107**
    Palinurus gilchristi **107**
    Panulirus argus **40**
    Panulirus cygnus **40**, **107**
    Panulirus ornatus **107**
    Sagmariasus verreauxi 27, 31, **33–35**, **37–40**, 60, **97**, **104**, *105*, **106**, 117–119, 132, 134, 141, 142, 153, **157**, 189, 220, 221
mud spiny lobster *see Panulirus polyphagus*
Museum of New Zealand Te Papa Tongarewa **42**, **43**, 45, **48**, **50**, 67, 121
musical furry lobster *see Palibythus magnificus*

Namibia **166–168**
Natal spiny lobster *see Palinurus delagoae*
National Institute of Water and Atmospheric Research Ltd (NIWA) 91, 94, 102, 121, 129, 133, 135, 201, 206

natural mortality rate (*M*) **176**
naupliosoma **86–89, 116, 133, 245**
navigation 37–39, **104, 105**
nervous system *also see* sensory system 29, **115, 218, 219**
New Plymouth **26**, 27, 34
New South Wales 34, **51, 60**, *61*, **89, 103, 105, 106, 121, 127, 156, 157, 172, 176, 178, 179**
New Zealand slipper lobster *see Antipodarctus aoteanus*
Nicobar Islands **68**
Ninety Mile Beach 25, **26, 99, 100**
nisto **95**
NIWA *see* National Institute of Water and Atmospheric Research Ltd
North Cape 25, **26**–28, **31, 33**, *35*, 49, 59. 98, 101, *152*, **154, 155, 193**, 221, 222
Nupalirus **53, 73, 76, 78, 92, 93**
Nupalirus chani 76, **78**
Nupalirus japonicus 76, **78**
Nupalirus vericeli 76, **78**
nursery 31, **33**
nutrition *also see* digestive system *and* prey **91, 94, 112, 124, 125, 200–205**

ocean acidification **197**
ocean climate *also see* El Niño-Southern Oscillation **166–170**, 177, 182, 183
octopus **108**, 123, 124, **135, 143**, 195
Oliver, Megan **206**
Oriental spear lobster *see Linuparus sordidus*
ornate spiny lobster *see Panulirus ornatus*
Otago **26, 36, 181–184**
output control 171, **174**, 197

packhorse *see Sagmariasus verreauxi*
packout *also see* exporting *and* marketing 138, **153, 155, 212–217**
painted spiny lobster *see Panulirus versicolor*
Palibythus **53, 73, 76, 78**, 92
Palibythus magnificus (musical furry lobster) 76, **78**
Palinurellus **53, 73, 76, 78, 92, 93**
Palinurellus gundlachi (furry lobster) 76, **78**
Palinurellus wieneckii (Indo-Pacific furry lobster) 76, **78**
Palinuridae **44**, 45, 47, **53, 55, 76, 77, 92, 93**
Palinurus 47, **50, 52, 53, 73, 76**, *79*, **92, 93**, *159*

Palinurus barbarae (Madagascar Ridge spiny lobster) **64**, 76, *79*
Palinurus charlestoni (Cape Verde spiny lobster) 76, *79*
Palinurus delagoae (Natal spiny lobster) **64**, 76, *79*, *107*, **160**
Palinurus elephas (European *or* common spiny lobster) 73, 76, *79*, **106, 127**, *159*, **160**, 199, **204, 205**
Palinurus gilchristi (southern spiny lobster) 76, *79*, *107*, *159*, **172**
Palinurus mauritanicus (pink spiny lobster) 76, *79*
Palinurus hügelii **48**, 49
Palinurus tumidus **48–50**
Palinurus verreauxi **46**, 47
Palinustus **53, 73, 76, 78, 92, 93**
Palinustus holthuisi (Asian blunthorn lobster) 76, **78**
Palinustus mossambicus (buffalo blunthorn lobster) 76, **78**
Palinustus truncatus (American blunthorn lobster) 76, **78**
Palinustus unicornutus (unicorn blunthorn lobster) 52, 76, **78**
Palinustus waguensis (Japanese blunthorn lobster) 76, **78**
Palliser Bay *26*, **147**
Panulirus **50**, 52, **53, 73, 77**, *80-82*, **92, 93, 130, 158**, *159*, **161, 202**
Panulirus argus (Caribbean spiny lobster) 76, 77
   age and growth **127, 136, 204**
   aquaculture **204**
   behaviour **135, 136**
   disease **136**
   distribution *80*, **158**, *159*
   enhancement **208**
   fishing and fishery **158**, *159*, **161, 162, 164, 165, 182, 195, 196**
   juvenile **40**, 123, **127, 208**
   movement/migration **40**
   puerulus **177–179**
Panulirus brunneiflagellum (aka-ebi) 77, *81*
Panulirus cygnus (western rock lobster) 77
   age and growth **204**
   aquaculture **204**
   autotomy **116**
   breeding 97, *107*, **128, 132, 172, 176**
   catch prediction **168, 169, 178–180**

distribution **80**
fishing and fishery **40**, **116**, **158**, *159*, **165**, 172, 178, **180**, **186**, 194
juvenile **40**, *107*, 158, 180
larval recruitment *107*
movement/migration **40**, *107*
ocean climate **168**, 169
puerulus *107*, **168**, 169, 178–180
Panulirus echinatus (brown spiny lobster) 77, **80**
Panulirus femoristriga (white-whiskered spiny lobster) **72**, 77, **80**
Panulirus gracilis (green spiny lobster) 77, **80**
Panulirus guttatus (spotted spiny lobster) 77, **81**, **128**, **132**
Panulirus homarus (scalloped spiny lobster) **74**, **76**, 77, **80**, **204**, **206**
Panulirus interruptus (California spiny lobster) 52, 77, **80**, **159**, **170**, **204**
Panulirus inflatus (blue spiny lobster) 77, **81**
Panulirus japonicus (Japanese spiny lobster) 77, **81**, *107*, *159*, **160**, **162**, **170**, **179**, 199, **202**, **208**, **209**, **211**
Panulirus laevicauda (smoothtail spiny lobster) 77, **82**, *159*, **182**
Panulirus longipes (longlegged spiny lobster) **76**, 77, **80**
Panulirus marginatus (Hawaiian *or* banded spiny lobster) 77, **80**, **170**
Panulirus ornatus (ornate spiny lobster) 77, **82**, **106**, *107*, *159*, **161**, 177, **202**, **204**, **205**
Panulirus pascuensis (Easter Island spiny lobster) 77, **82**, **110**
Panulirus penicillatus (pronghorn spiny lobster) 77, **81**
Panulirus polyphagus (mud spiny lobster) 77, **81**
Panulirus regius (royal spiny lobster) 77, **80**
Panulirus stimpsoni (Chinese spiny lobster) 77, **82**
Panulirus versicolor (painted spiny lobster) 77, **82**
Papua New Guinea 99–101, *159*
parasite *also see* disease **136**
Parengarenga Harbour **25**, *26*, 59
pathogen *see* disease
Pauly, Daniel **152**, 192
PaV1 virus **136**
pawharu 43, **150**
pesquero **161**, **162**, **208**
pheromone **134**, **135**
Phillips, Bruce 161, **168**, 210

phyllamphion **92**
Phyllosoma **92**
phyllosoma 85–94, *104*, *105*, **106**, *107*, 116, 123, **170**, **197**, 198, **200**, **201**, **204**, 245
Antipodarctus **93**
Arctides **93**
behaviour 85, **91**, **94**, **102**–**104**, **106**, *107*, **118**, **200**, **201**, **203**
culture 91, **199**–**204**
Ibacus **93**
identification **92**, **93**
Jasus 68, **92**, **93**
Jasus edwardsii **86**, **88**, 97, **106**, *107*, 198, 200, 201
Jasus lalandii 97, 199, **200**
Justitia **92**, **93**
Linuparus **92**
Nupalirus **92**, **93**
Palinurus **92**, **93**
Palinurus elephas **106**, **204**, **205**
Palinurellus **92**, **93**
Palinustus **92**, **93**
Panulirus **92**, **93**
Panulirus cygnus 97, *107*
Panulirus homarus 204
Panulirus ornatus **106**, *107*, 204
phyllamphion **92**
Puerulus **92**, **93**
Sagmariasus **92**, **93**
Sagmariasus verreauxi **85**–**94**, **96**–**102**, *103*–*105*, **117**, **118**, **199**–**204**
scyllarid **92**, **93**
Scyllarides **93**
phylogeny and evolution **53**–**58**, **116**–**119**, 130
Pike, Richard **42**, **43**, **86**–**88**
pink spiny lobster *see Palinurus mauritanicus*
Pollock, Dave **56**
polystyrene **156**, **216**, **217**
Poor Knights Islands **152**, **154**
postlarva *see* puerulus and puerulus settlement 245
potting *also see* fishing and fisheries 28, 29, 33, 38, 39, **64**, **65**, **68**–**71**, **137**–**160**, **164**, **171**, **175**–**177**, 192, **193**–**196**, 213, 215, 220
pouraka 149
prawn killer *see Ibacus alticrenatus*
predator 40, **55**, 84, 85, **91**, **105**, **108**, **117**, **122**–**124**, 126, **135**, **143**, **170**, **196**, **206**–**208**

predicting catches *see* catch prediction
pre-recruit 177, 178, 180, 183, **246**
prey *also see* nutrition 91, 94, 102, 124–126, 201–203
processing 138, 149, 153, 155, 156, 212–217
production hypothesis 163
Projasus 53, 55, 58, 73, 74, 77, *78*, 92, 95, 159
Projasus bahamondei (Chilean jagged lobster) 64, 73, 77, *78*, *159*
Projasus parkeri (Cape jagged *or* deepwater lobster) 53, 55, **64, 69, 70, 72**, 74, 77, *78*, *159*
pronghorn spiny lobster *see Panulirus penicillatus*
Puerulus 52, **53**, 73, 74, 77, *83*, 92, 93, *159*
Puerulus angulatus (banded whip lobster) 77, *83*
Puerulus carinatus (red whip lobster) 77, *83*
Puerulus sewelli (Arabian whip lobster) 77, *83*, *159*
Puerulus velutinus (velvet whip lobster) 77, *83*
puerulus and puerulus settlement 86, 89, 90, 95, **104, 105, 106, 122, 123**, 126, 170, 177–179, 189, 202, 205, **246**
  Jasus edwardsii 95, **105, 106**, *107*, **108**, 118, **126, 167, 168, 170, 178, 179, 181, 183–186, 207**
  Panulirus argus 178–179
  Panulirus cygnus *107*, 159, **168, 169**, 178–180, 186
  Panulirus japonicus *107*, **170, 208, 209**
  Panulirus ornatus *107*, **205**
  Sagmariasus verreauxi 33, 34, 89, 90, 94–96, 102, 103, *104*, **105, 106**, 118, **121, 122, 126**, 178, 179, **200–203**
puerulus collector 89, **121, 177–179**, 205
Puerulus sewelli (Arabian whip lobster) 77, *83*, *159*, 162
Pukaki 28, 29

quota 144, 157, **174**, 197, 220, 221
Quota Management System (QMS) **144**

Raethke, Natalie **135**
recreational fishing **163–165**, 174, 181
  Panulirus argus **164**, 165
  Panulirus cygnus 165
  Sagmariasus verreauxi 154, **156**, **163–165**
recruitment **246**

larval/postlarval 86, **96–103**, *104*, *105*, 106, *107*, 117, 118, **168–170**, 177–180, 181, **183–186**, 188, **189**
  mechanism 86, **96–103**, *104*, *105*, 106, *107*, **168–170**
  to the breeding population 86, *104*, *105*, *107*, 118, **172–174**,
  to the fishery **167–170, 177–180, 183–186**
recruitment overfishing **177**, 178, 180, **182**, 220, **246**
red rock lobster *see Jasus edwardsii*
red whip lobster *see Puerulus carinatus*
regeneration of limbs **142**
regulations *also see* management 27, 153, **156, 157, 163, 164**, 166–168, **171–174**, 196
reproduction *see* breeding
respiratory system **113**
restocking **206–209**
Réunion Island **65–68**
Ritar, Arthur **202**
royal spiny lobster *see Panulirus regius*
Russell 67, 121, **137, 138**, 144

Sagmariasus 51, **53, 55, 56**, 58, 73, 77, 92, 93
Sagmariasus flemingi 54, 55
Sagmariasus verreauxi (packhorse spiny lobster *or* eastern rock lobster) 77
  age and growth **117, 126–128, 142**, 202, **204**
  aquaculture 198, **199–204**
  Australia 34, 51, **56, 60, 61**, 87, **103, 104**, *105*, 106, 119, **156, 157, 165, 171, 172**, 178, 179, 199, **202, 203**
  autotomy **116**
  breeding 33, **39**, 59, **60**, 87, 88, **97, 103**, *104*, *105*, **117, 118, 127, 128, 130, 132, 133**, 173, 200, **201**
  bycatch **193**, 194
  disease **135**, 136
  distribution 31, **33**, *59–61*
  egg laying, fertilisation, development and hatching **39**, 87, 88, **97, 116–118, 130**, **201**
  fecundity 87, **117, 132**
  fishing and fishery **25–28**, 31, **33**, 61, **137–157**, 161, **163–165**, 171, 172, 176, **193**, 194, 220–222
  form 32, **43, 44**, 72, **110, 111, 112, 117, 118**

Hopkins, Bill 39, **143–146**, **155**, 176
juvenile **27–33**, *35*, 37, **44**, **60**, **61**, *104*, *105*, 106, **117**, **121**, 122, **126**, **154**, 187, **188**, **200–203**
Kittaka, Jiro 89, **94**, **96**, **199**, **200**
larval recruitment **96–103**, *104*, *105*, **106**
management and regulations 27, **156**, **157**, **171–174**, 220–223
Māori fishing **147–151**
marine protected area **189**, **190**
movement/migration **27–35**, **37–40**, 97, *104*, *105*, **106**, **117–119**, 132, 134, 141, 142, 153, 157, 189, 190, 220, 221
naupliosoma **87–89**, 116, 133
nursery **31**, 33, *104*, *105*
phyllosoma, and larval culture 33, **86–94**, **97–103**, *104*, *105*, **106**, 118, **199–203**
phylogeny and evolution **53–58**
processing and marketing 138, **153**, **155**, **156**, **212**, **213**, **215**, **218**, **219**
puerulus and puculus settlement 34, **89**, **90**, **94–96**, **102**, **103**, *104*, *105*, **106**, 117, 118, **121**, **122**, **126**, **178**, **179**, **199–201**, **203**
recruitment and recruitment mechanism 34, **96–103**, *104*, *105*, **106**
regeneration of limbs **116**, **142**
size and age at onset of breeding **117**, **127**, **128**, **171**, **172**
tagging **29–31**, 39, 142
taxonomy and naming **45–51**
Samuels, Dover **149–151**
scalloped spiny lobster *see Panulirus homarus*
Scyllaridae/slipper lobsters **44**, **47**, **55**, 58, **74**, **75**, **92**, **93**, **95**, **124**, 245, **246**
Scyllarides haanii **75**, **93**
seamount **44**, **61–65**, **70**, **71**, 72, **74**, **246**
sensory system *also see* nervous system **38**, **39**, **115**, **203**, **218**, **219**
settlement *see* puerulus and puerulus settlement
shell *see* exoskeleton
shelter fidelity **134**
shifting baseline syndrome **152**, 192
shrimp **95**
Silentes **50**, **53**, **55**, **64**, **76**, **247**
size at onset of breeding (SOB) 37, **117**, **127–129**, 171, **172**, **196**, **246**
size composition 142, **147**, **154**, **176**, 177, 181, **188**, **246**

slipper lobster *see* Scyllaridae
smoothtail spiny lobster *see Panulirus laevicauda*
social behaviour *see* behaviour
South Africa **40**, **41**, 69, 97, *107*, **158**, *159*, **160**, 168, 172, **176**, 181, 199
South Atlantic Ocean **71**
South Pacific Ocean **56**, **57**, 60, **62–64**, 67, **70**, **71**, **74**, 80, **100**, **110**
Southern Oscillation Index (SOI) *also see* El Niño-Southern Oscillation *and* ocean climate **168–170**, 180, 183, **247**
Southern Raider **65–70**
southern rock lobster *see Jasus edwardsii*
southern spiny lobster *see Palinurus gilchristi*
Southland Current **35**, *36*, **98**
Spanish lobster *see Arctides antipodarum*
spawning *see* breeding (egg laying) **247**
spawning stock **157**, **172–174**, **176**, **189**, **247**
Spirits Bay **26**, 96, **128**, 137, 139, **144**, **145**, **154**, **155**, 187, 194
spotted spiny lobster *see Panulirus guttatus*
St. Paul Island **62**, **65**, **66**, **68**, **69**
St. Paul rock lobster *see Jasus paulensis*
statocyst **115**
Stewart Island **26**, **35**, **36**, **60**, *61*, 70, **183**
stock assessment 41, **174–186**, 220, **247**
Street, Bob **35**, 37
Stridentes **50**, **53**, **55**, **76**, **135**, **247**
stridulating organ **50**, **53**, **55**, **135**, **247**
Sykes, Daryl **121**, **216**
Synaxidae **53**

tagging **29**, **30**, 39, 142
tail fan necrosis **135**, **136**, **196**
tailing *also see* processing 138, **153**, **212**, **213**, **217**, **219**
tangle net **160**, 162
taruke **147**, 149
Tasman Sea **60**, *61*, **64**, **65**, **103**, **104**
Tasmania *61*, **105**, **178**, **193**, 202, 203, **207**, **218**
taste **149**, **202**, **211**, **217**, 218
taxonomy **45–47**
Te Papa *see* Museum of New Zealand Te Papa Tongarewa
Te Tapuwae o Rongokako Marine Reserve **190**, **191**, **195**, **196**
Three Kings Islands *26*, 96, **99**, **100**, 152
Tom Bowling Bay **26** 33, 39

Tong, Len **94**, 198, **201**
Torres Strait *107*, ***159***, **161**, **177**
total allowable catch (TAC) **174**
total allowable commercial catch (TACC) 144, **155**, **174**, 184, 220, 221
total mortality rate (Z) **176**, **177**
trammel net *see* tangle net
translocation **206–208**
trapping *see* potting
trawling **36**, 70, 145, 153, **155**, *159*, **160–162**, 195
Tristan da Cunha 61, ***62***, **63**
Tristan rock lobster *see Jasus tristani*
trophic cascade **125**
turgid lobster syndrome **136**

unicorn blunthorn lobster *see Palinustus unicornutus*
unintended catches *see* bycatch

velvet whip lobster *see Puerulus velutinus*
Vema Seamount ***62***, **63**
Verreaux, Jules Pierre **47**
Vietnam **205**
viral disease **136**, 201
von Hügel, Karl Alexander Anselm Freiherr **48**

Wanganella Bank ***61***
Webber, Rick **45**, **50**, **70**, **71**, 88, **92**, 198
West Auckland Current ***98***, **99**, **101–103**
Western Australia **40**, 97, ***107***, **116**, **128**, **158**, **159**, **165**, **168**, **169**, **172**, **176**, **178**, **180**, **186**, **194**
western rock lobster *see Panulirus cygnus*
Westley, Scott **157**
Whangaroa *26*, **48**, **49**, 59, *152*, **153**
White Island *26*, 33, 59, *152*, **154**
white-whiskered spiny lobster *see Panulirus femoristriga*